THE FUTURE OF

The Future of Warfare in 2030

Project Overview and Conclusions

RAPHAEL S. COHEN, NATHAN CHANDLER, SHIRA EFRON,
BRYAN FREDERICK, EUGENIU HAN, KURT KLEIN,
FORREST E. MORGAN, ASHLEY L. RHOADES,
HOWARD J. SHATZ, AND YULIYA SHOKH

Prepared for the United States Air Force
Approved for public release; distribution unlimited

For more information on this publication, visit www.rand.org/t/RR2849z1

Library of Congress Cataloging-in-Publication Data is available for this publication.
ISBN: 978-1-9774-0295-0

Published by the RAND Corporation, Santa Monica, Calif.

Cover, spine: combo1982/Getty Images, matejmo/Getty Images, StudioM1/Getty Images

Preface

Where will the next war occur? Who will fight in it? Why will it occur? How will it be fought? Researchers with RAND Project AIR FORCE's Strategy and Doctrine Program attempted to answer these questions about the future of warfare—specifically, those conflicts that will drive a U.S. and U.S. Air Force response—by examining the key geopolitical, economic, environmental, geographic, legal, informational, and military trends that will shape the contours of conflict between now and 2030. This report is one of a series that grew out of this effort. The other reports in this series are

- Raphael S. Cohen, Eugeniu Han, and Ashley L. Rhoades, *Geopolitical Trends and the Future of Warfare: The Changing Global Environment and Its Implications for the U.S. Air Force* (RR-2849/2-AF)
- Forrest E. Morgan and Raphael S. Cohen, *Military Trends and the Future of Warfare: The Changing Global Environment and Its Implications for the U.S. Air Force* (RR-2849/3-AF)
- Howard J. Shatz and Nathan Chandler, *Global Economic Trends and the Future of Warfare: The Changing Global Environment and Its Implications for the U.S. Air Force* (RR-2849/4-AF)
- Shira Efron, Kurt Klein, and Raphael S. Cohen, *Environment, Geography, and the Future of Warfare: The Changing Global Environment and Its Implications for the U.S. Air Force* (RR-2849/5-AF)
- Bryan Frederick and Nathan Chandler, *Restraint and the Future of Warfare: The Changing Global Environment and Its Implications for the U.S. Air Force* (RR-2849/6-AF).

This volume introduces the overall definitions, methodology, findings and the larger implications for the Air Force in particular and the joint force at large that are used in the series of reports.

This research was sponsored by the Director of Strategy, Concepts and Assessments, Deputy Chief of Staff for Strategic Plans and Requirements (AF/A5S). It is part of a larger study, entitled *The Future of Warfare*, that assists the Air Force in assessing trends in the future strategic environment for the next Air Force strategy. This report should be of value to the national security community and interested members of the general public, especially those with an interest in how global trends will affect the conduct of warfare. Comments are welcome and should be sent to the project leader, Raphael S. Cohen. Research was completed in August 2018.

RAND Project AIR FORCE

RAND Project AIR FORCE (PAF), a division of the RAND Corporation, is the U.S. Air Force's federally funded research and development center for studies and analyses. PAF provides the Air Force with independent analyses of policy alternatives affecting the development, employment, combat readiness, and support of current and future air, space, and cyber forces. Research is conducted in four programs: Force Modernization and Employment; Manpower, Personnel, and Training; Resource Management; and Strategy and Doctrine. The research reported here was prepared under contract FA7014-16-D-1000.

Additional information about PAF is available on our website: www.rand.org/paf.

This report documents work originally shared with the U.S. Air Force in September 2018. The draft report, issued September 7, 2018, was reviewed by formal peer reviewers and U.S. Air Force subject-matter experts.

Contents

Figures and Tables

Figures

Tables

Summary

The U.S. track record for predicting the future of warfare is notoriously poor. Robert Gates, U.S. Secretary of Defense from 2006 to 2011, famously quipped, "When it comes to predicting the nature and location of our next military engagements, since Vietnam, our record has been perfect. We have never once gotten it right, from the Mayaguez to Grenada, Panama, Somalia, the Balkans, Haiti, Kuwait, Iraq, and more—we had no idea a year before any of these missions that we would be so engaged."[1] And yet, for better or worse, the U.S. military is deeply invested in the forecasting business because the services need to start building today what will be needed one or even two decades from now. Thus, the question becomes how to predict the future of warfare correctly.

In this report, we start by exploring why the U.S. military so often fails to the predict the future correctly and find that failed predictions cannot be chalked up simply to stupidity of individual leaders, ignorance of technology, or failure to identify trends. Rather, the failures stem from not thinking comprehensively about the factors that shape conflict and how these variables interact with one another. We then develop definitions and a methodology to do just that—first identifying the key three dozen or so geopolitical, economic, environmental, geographic, military, legal, and informational trends that will shape the future of warfare from now until 2030 and then aggregating them to paint a holistic picture of which countries the United States

[1] Robert Gates, speech to the U.S. Military Academy, West Point, N.Y., February 25, 2011.

will fight with and against, where these conflicts will occur, what they might look like, how the United States will wage them, and when and why the United States might go to war in the first place.

Based on this analysis and assuming that the United States will continue to try to maintain its position as the world's only global military superpower, we conclude that the United States will confront a series of deepening strategic dilemmas when confronting warfare from now through 2030. U.S. adversaries—China, Russia, Iran, North Korea, and terrorist groups—likely will remain constant, but U.S. allies are liable to change as Europe becomes increasingly fragmented and inward-looking and as Asia reacts to the rise of China. The locations where the United States is most likely to fight will not match where conflicts could be most dangerous to U.S. interests. The joint force will face at least four diverse types of conflict, each requiring a somewhat different suite of capabilities, just as it confronts diminishing quantitative and qualitative military advantages. Above all, perhaps, the United States of 2030 could progressively lose the capacity to dictate strategic outcomes and to shape when and why the wars of the future occur.

Ultimately, as the future of warfare places more demands on U.S. forces and pulls limited U.S. resources in opposite directions, the United States will face a grand strategic choice: It can break with the internationalist foreign policy that it has pursued since at least the end of the Cold War and become dramatically more selective about where, when, and why it commits forces. Alternatively, it can double down on its commitments, knowing full well that doing so will come with significantly greater cost—in treasure and perhaps in blood. From the narrower perspective of the joint force and U.S. Air Force, the future of warfare will require additional capability investments in precision, information, and automation; more capacity to tackle the challenges posed by each of the five aforementioned U.S. adversaries; enhanced forward posture in all three theaters of concern (including the Middle East); and a renewed emphasis on agility at every echelon.

Acknowledgments

This study would not have been possible without the help of many people. First and foremost, we thank Brig Gen David Hicks, Col Linc Bonner, and Scott Wheeler of the Air Force A5S for sponsoring this study and guiding it along the way. We would also like to thank Paula Thornhill, the Project AIR FORCE strategy doctrine program director, for her guidance and mentorship of this study. We also thank Thomas Mahnken and Michael Mazarr, who reviewed an earlier draft of this work and who helped improve the manuscript considerably. We also owe special thanks to Arwen Bicknell for her expert editing of this volume and the others in the series. Last but not least, the research team owes a special debt of gratitude to more than a hundred experts across the globe who volunteered their time to give their perspectives on the future of warfare both within their region and globally. Human subjects protocol prevents us from thanking individuals by name, but we would like to thank the following institutions for hosting our research visits: United Kingdom (UK) Ministry of Defence's Development, Concepts, and Doctrine Centre; UK Parliament; Royal Institute of International Affairs (Chatham House); Control Risks; Royal United Services Institute for Defence and Security Studies; London School of Economics and Political Science IDEAS; International Institute of Strategic Studies; North Atlantic Treaty Organization Headquarters; European Commission; Centre for European Policy Studies; European Centre for International Political Economy; German Federal Ministry of Defence; German Institute for International and Security Affairs; German Federal Parliament; German Federal Foreign Office;

European Council on Foreign Relations; Polish Ministry of Foreign Affairs; Polish Ministry of National Defence; Polish Institute of International Affairs; Centre for Eastern Studies; Interdisciplinary Center Herzliya; University of Jordan; Jordanian Army Forces; Royal Jordanian Air Force; Strategic Intelligence Solutions; Middle East Media and Policy Studies Institute–Centre for Strategic and International Studies; Strategic Intelligence Solutions; U.S. Embassy, Amman, Jordan; U.S. Embassy, Abu Dhabi, United Arab Emirates; Peking University Institute for International and Strategic Studies; Institute for China–U.S. People to People Exchange; Chinese Academy of Social Sciences; China Institute for Contemporary International Relations; University of International Relations; China Institute for International Studies; China Institute for International Strategic Studies; China Foundation for International Strategic Studies; Carnegie Endowment/Tsinghua University; Agence France Presse; *New York Times*; *Wall Street Journal*; *China Policy*; Japanese Ministry of Defense; Japanese Ministry of Foreign Affairs; Sasakawa Peace Foundation; and Hosei University.

Abbreviations

AI	artificial intelligence
DoD	U.S. Department of Defense
GDP	gross domestic product
ISR	intelligence, surveillance, and reconnaissance
MENA	Middle East and North Africa
NATO	North Atlantic Treaty Organization
RMA	Revolution in Military Affairs
USAF	United States Air Force

CHAPTER ONE

The Future of Warfare

"When it comes to predicting the nature and location of our next military engagements, since Vietnam, our record has been perfect. We have never once gotten it right, from the Mayaguez to Grenada, Panama, Somalia, the Balkans, Haiti, Kuwait, Iraq, and more—we had no idea a year before any of these missions that we would be so engaged."

—U.S. Secretary of Defense Robert Gates[1]

Of all people, Robert Gates, a lifelong intelligence officer who rose to become the Director of Central Intelligence before becoming Secretary of Defense, should know a thing or two about forecasting the future of warfare. After all, he made a career of being both a producer and consumer of these predictions, so his admonition that analysts have consistently failed for decades to predict where, why, and how U.S. forces will be employed even a year later should be heeded. And yet, for better or worse, the U.S. military is deeply invested in the forecasting business. All the armed services regularly take gambles on what they will need to win tomorrow's wars because both technologies and people take decades to develop. Thus, the challenge becomes to define the future of conflict with enough clarity to be useful for military planning while remaining mindful of Gates' warning about the limitations of the practice.

[1] Robert Gates, speech to the U.S. Military Academy, West Point, N.Y., February 25, 2011.

In this report, we summarize the findings of a RAND Corporation research project that made one such attempt. The project produced a series of reports, each dedicated to how certain key trends will shape conflict between now and 2030 and what this might mean for the U.S. Air Force (USAF), the joint force, and the United States at large. This report—the first in the series—provides an overview of the project's methodology and key findings and of the implications for the USAF and joint force.[2] First, we ask why militaries so often fail to correctly predict the future of conflict, and we argue that these lapses often stem from failing to think holistically about the changes in environment. Second, we situate our study within the context of similar efforts and build on that previous work to develop definitions of *the future* and of *warfare*. Third, we describe our methodology and summarize three dozen or so geopolitical, economic, environmental, geographic, military, legal, and informational trends that will shape the future of warfare from now until 2030. (These trends are explored in depth in the companion volumes of this report.) Fourth, we draw on these trends to paint a holistic picture of the future of warfare—the potential allies and enemies of the United States, where conflicts will occur, what they might look like, how the United States will wage them, and when and why the United States might go to war in the first place. Finally, we conclude by describing the implications of this work for the USAF and the joint force.

---

[2] The other reports in this series are Raphael S. Cohen, Eugeniu Han, and Ashley L. Rhoades, *Geopolitical Trends and the Future of Warfare: The Changing Global Environment and Its Implications for the U.S. Air Force*, Santa Monica, Calif.: RAND Corporation, RR-2849/2, 2020; Forrest E. Morgan and Raphael S. Cohen, *Military Trends and the Future of Warfare: The Changing Global Environment and Its Implications for the U.S. Air Force*, Santa Monica, Calif.: RAND Corporation, RR-2849/3, 2020; Howard J. Shatz and Nathan Chandler, *Global Economic Trends and the Future of Warfare: The Changing Global Environment and Its Implications for the U.S. Air Force*, Santa Monica, Calif: RAND Corporation, RR-2849/4, 2020; Shira Efron, Kurt Klein, and Raphael S. Cohen, *Environment, Geography, and the Future of Warfare: The Changing Global Environment and Its Implications for the U.S. Air Force*, Santa Monica, Calif.: RAND Corporation, RR-2849/5, 2020; and Bryan Frederick and Nathan Chandler, *Restraint and the Future of Warfare: The Changing Global Environment and Its Implications for the U.S. Air Force,* Santa Monica, Calif.: RAND Corporation, RR-2849/6, 2020.

Ultimately, we conclude that if the United States wants to maintain its position as the world's only global military superpower, it will confront a series of deepening strategic dilemmas when confronting the future of warfare through 2030. U.S. adversaries—China, Russia, Iran, North Korea, and terrorist groups—will likely remain constant, but U.S. allies are liable to change as Europe becomes increasingly fragmented and inward-looking and as Asia reacts to the rise of China. The locations where the United States is most likely to fight will not match where conflicts could be most dangerous to U.S. interests. The joint force will face at least four diverse types of conflict, each requiring a somewhat different suite of capabilities, just as it confronts diminishing quantitative and qualitative military advantages. Above all, perhaps, the United States of 2030 could progressively lose the capacity to dictate strategic outcomes and to shape when and why the wars of the future occur.

CHAPTER TWO

The Failures of Forecasting the Future

Military history is littered with mistaken predictions about the future of warfare that left the forecasters militarily unprepared—sometimes disastrously so—for the conflict ahead; indeed, the 20th century alone features several debacles. On the eve of World War I, German Chancellor Theobald von Bethmann Hollweg claimed that the wars of the future would be "decisive" and a "brief storm."[1] The Great War's four years and millions of casualties proved Hollweg and many others wrong, but it did not make European militaries any better at predictions. During the interwar period, France bet big with the Maginot Line—a sophisticated but largely ineffectual series of fortifications built along the French-German border—that the next war would also center on the largely static trench warfare of the Great War. This miscalculation contributed to the country's defeat in a mere six weeks in May–June 1940.[2]

The United States has also suffered its share of bad predictions. Prior to the Vietnam War, the U.S. military assumed that the next big war would be large-scale conventional or nuclear conflict against the Soviet Union in Europe, not irregular conflict in Asia; this helped pave the way to the most searing defeat in U.S. history.[3] More recently,

[1] Stephen Van Evera, "The Cult of the Offensive and the Origins of the First World War," *International Security*, Vol. 9, No. 1, Summer, 1984.

[2] For a concise analysis of why France failed to anticipate and adapt to the German threat, see Eliot A. Cohen and John Gooch, *Military Misfortunes: The Anatomy of Failure in War*, New York: Free Press, 1990, pp. 197–230.

[3] In his classic study of the U.S. Army in Vietnam, Andrew Krepinevich attributes the Army's mistaken conceptions about the future of warfare and "how wars *ought* to be waged,"

even after the U.S. invasion of Afghanistan, then–Secretary of Defense Donald Rumsfeld argued that the future of warfare belonged to the Revolution in Military Affairs (RMA)—a concept in which information, speed, precision, and range could substitute for mass and firepower—and not to the manpower-intensive counterinsurgency campaigns that ensued over the subsequent decade.[4]

Why do predictions about the future of warfare so often fall flat? Mistaken forecasts about the future of conflict are not unique to any one state or relegated to any particular period. As the examples demonstrate, the German, French, and U.S. militaries all have fallen prey to bad predictions despite being among the most sophisticated militaries of their times. And this is by no means an exhaustive list of even 20th-century examples.[5] Perhaps no country has ever gotten the future of warfare entirely right.

Similarly, bad predictions cannot be chalked up simply to the folly of individual decisionmakers. For example, British military historian Michael Howard argues that European military leaders' faith in the offensive on the eve of Great War did not stem from ignorance or naiveté: "Nobody was under any illusion, even in 1900, that frontal attack would be anything but very difficult and that success could be purchased with anything short of very heavy casualties."[6] Rather, Howard claims, lessons from the Russo-Japanese War and other conflicts taught European leaders that "[w]ar had been shown to be neither impossible, nor suicidal. It was still a highly effective instrument of policy for a nation which had the courage to face its dangers and the endurance to bear its costs—especially its inevitable and predictable

---

which he terms the "Army Concept." See Andrew F. Krepinevich, Jr., *The Army and Vietnam*, Baltimore, Md.: Johns Hopkins University Press, 1986, p. 5.

4 See Donald H. Rumsfeld, "Transforming the Military," *Foreign Affairs*, Vol. 81, No. 3, May–June 2002.

5 Other prominent examples of other first-rate military powers failing to understand the future of warfare include Britain and the Battle of Gallipoli, U.S. failures during the Korean War, and Israel during the 1973 Arab-Israeli conflict, among many others. See Cohen and Gooch, 1990.

6 Michael Howard, "Men Against Fire: Expectations of War in 1914," *International Security*, Vol. 9, No. 1, Summer, 1984, p. 43.

costs in human lives."[7] The same could also be said of French and U.S. blunders: On a certain level, French ideas about static defenses and the U.S. focus on large-scale European conflict during the Cold War and on the RMA in the 1990s were entirely logical, grounded in recent experiences, and possibly even partially correct.[8] The U.S. focus on winning a large-scale European conflict, for example, possibly deterred such a conflict from ever occurring. But we can see only the Vietnams, not the counterfactual wars that never happened.

Bad predictions are also rarely as simple as turning a blind eye to military technological advances. Despite missing the revolution in mobile armored warfare, France still developed and fielded the *Char B*, arguably the best tank on the Western front in May 1940.[9] The U.S. Army still developed the workhorse of Vietnam, the helicopter—just not the counterinsurgency doctrine to go along with it.[10] And, if anything, Rumsfeld's ideas about the RMA stemmed from being overly enamored with technology—specifically information and airpower—rather than a lack of technological awareness.

Likewise, poor forecasts are not the result of a simple inability to correctly identify trends. According to a recent study of post–Cold War U.S. defense, most analyses are actually correct about most of the broad, long-term trends they identify. Of 66 defined predictions about the future made across ten major defense reviews, 40 were correct and another eight were partially correct.[11] The problem with these reviews was that the trends they identified were based on an incomplete world picture that was missing key events,

---

[7] Howard, 1984, p. 47.

[8] France's assumptions were based on hard-learned lessons during World War I. U.S. assumptions about the next big war being in Europe came from its experiences during World War II, not to mention actual Soviet Bloc force posture. Finally, the RMA was grounded in seeming lessons from Operation Desert Storm in the Middle East and, later, Operation Allied Force in Kosovo.

[9] R. H. S. Stolfi, "Equipment for Victory in France in 1940," *History*, Vol. 55, No. 183, February 1970, p. 2.

[10] For a detailed discussion, see Krepinevich, 1986, pp. 112–127.

[11] From Raphael S. Cohen, *The History and Politics of Defense Reviews*, Santa Monica, Calif.: RAND Corporation, RR 2278-AF, 2018, p. 55.

such as those that reshaped the nature of warfare in the September 11, 2001, terrorist attacks (and the subsequent Global War on Terrorism), the second Iraq War (and the need to mount a counterinsurgency effort), the Arab Spring (and the rise of the Islamic State), or Russia's aggression in Ukraine (and the return of great-power security challenges to Europe).[12]

Finally, forecasting is not simply an impossible task. The United States, despite its recent lackluster track record, has predicted conflicts successfully in the past. For example, in December 1921, the Joint Planning Committee concluded, "It may safely be assumed that Japan is the most probable enemy"; that Japan would eventually militarily challenge U.S. interests in the Far East, particularly claims in Guam and the Philippines; and that the United States would have to launch an offensive war to reclaim likely losses in the initial phases of the conflict.[13] Although U.S. planners did not predict when and how war would break out, they did successfully develop War Plan Orange and some of the basic concepts of "island-hopping" and economic warfare that would be employed to great effect decades later during World War II.[14]

More often, poor predictions stem from failing to think holistically about the factors that drive changes in environment and their implications for warfare. Part of this, of course, comes down to how advances in technology might alter the way force can be employed on the battlefield. Indeed, military organizations historically have struggled to understand and embrace new technologies (especially those that challenge their core identities), and there is vigorous academic debate about whether militaries require outside civilian pressure to do so.[15]

[12] Cohen, 2018, p. 56.

[13] Louis Morton, "War Plan Orange: Evolution of a Strategy," *World Politics*, Vol. 11, No. 221, 1959, p. 228.

[14] Morton, 1959, pp. 232–233.

[15] For an example of this debate, see Barry R. Posen, *The Sources of Military Doctrine: France, Britain, and Germany Between the World Wars*, Ithaca, N.Y.: Cornell University Press, 1984; Stephen Peter Rosen, "New Ways of War: Understanding Military Innovation," *International Security*, Vol. 13, No. 1, Summer, 1988; Williamson R. Murray and Allan R. Millett, eds., *Military Innovation During the Interwar Period*, Cambridge, UK: Cambridge University Press, 1998.

But comprehensive thinking about the future of warfare needs to extend beyond just understanding the operational implications of technology. After all, geopolitical changes—rather than technological ones—pushed the U.S. military from preparing for high-end conflict against a peer competitor during the late Cold War to combating sub-state instability in Somalia, Haiti, and the Balkans in the 1990s and fighting radical Islamic terrorism during the 2000s, then back to once again preparing for high-end, near-peer conflict today. Others argue that environmental changes—as much as political factors—are often key drivers of conflict, with the Syrian civil war a notable example.[16] Economic factors also have long been considered to influence conflict, but in an increasingly globalized world in which a nation's lifeblood can depend on trade and access to international credit, economic actions also can serve as powerful tools of warfare used in lieu of violence. Still other factors—such as international laws, public opinion, and media coverage—can constrain the way that states use force—and, consequently, how wars are fought.[17]

Successful forecasting efforts reflect a wide-ranging examination of many of these different variables. War Plan Orange, for example, considered geographic, political, and economic variables along with military considerations.[18] The inherent challenge in forecasting comes from trying to combine all these factors into a coherent picture of the future.

---

[16] For example, see Peter H. Gleick, "Water, Drought, Climate Change, and Conflict in Syria," *Weather, Climate, and Society*, Vol. 6, No. 3, July 2014.

[17] For seminal work on how international norms shape conflict, see Peter J. Katzenstein, ed., *The Culture of National Security: Norms and Identity in World Politics*, New York: Columbia University Press, 1996.

[18] Indeed, as early as 1919, U.S. Navy CAPT Harry Yarnell—one of the War Plan Orange planners—argued that any military plan needed to address questions about U.S. objectives and economic interests that fell outside military lanes, stating that "[t]hese questions are not for the War and Navy Departments to answer, but for the State Department." Morton, 1959, p. 225.

CHAPTER THREE

Studying the Future Today

The challenges of predicting the future of warfare have not prevented plenty from trying. Although not focused on warfare per se, the National Intelligence Council releases an analysis of long-term global trends every four years and is now on its sixth edition of the series.[1] Within the U.S. Department of Defense (DoD), the Joint Staff published its attempt to forecast the future of warfare in *Joint Operating Environment (JOE) 2035: The Jointed Force in a Contested World* in 2016.[2] The services—including the USAF—all publish their own long-term forecasts of the future of warfare from their perspectives.[3] Foreign militaries commonly publish their takes on how warfare will evolve over the decades.[4] There is also vigorous academic discussion of the topic.

These reports vary in length and depth, and each has its own unique flavor, but there are common elements that run across them. On the most basic level, all the government-sponsored analyses tend to touch on the same broad categories—political, military, economic,

---

1 National Intelligence Council, *Paradox of Progress*, Washington, D.C., Global Trends main report, January 2017, p. vi.

2 Joint Chiefs of Staff, *Joint Operating Environment (JOE) 2035: The Joint Force in a Contested World*, Washington, D.C., July 14, 2016.

3 For example, see Department of the Air Force, *Air Force Strategic Environment Assessment 2014–2034*, Washington, D.C., 2015; U.S. Marine Corps, Futures Directorate, *2015 Marine Corps Security Environment Forecast: Future 2030–2045*, Quantico, Va., 2015.

4 For example, see Modernisation and Strategic Planning Division, Australian Army Headquarters, *Future Land Warfare*, Canberra, 2014; Ministry of Defence UK, *Strategic Trends Programme: Future Operating Environment 2035*, London, November 30, 2014.

environmental, geographic, legal, and informational trends—and try to paint a comprehensive future conflict. Academic studies, by contrast, often are more limited in scope and more tailored in their approach, extrapolating from a single war or category of trend.[5]

Many of these studies also agree on the relevant trends. For example, there is broad agreement that global, if uneven, growth and increasing urbanization will define how societies are structured and by extension, how conflict will be waged.[6] These studies also often focus on climate change and its second-order impacts on the food supply and access to natural resources as potential causes of conflict.[7] Similarly, most of these studies note that increasing the political, economic, and informational effects of globalization creates potential benefits and vulnerabilities for national security as it becomes easier for threats to transcend borders.[8] Overall, there tends to be widespread agreement that the future looks bleak—filled with both interstate and intrastate tension—and that it will become harder for the United States and its allies to wage wars successfully.

Viewed as a whole, however, this panoply of documents has two limitations. First, these studies struggle with balancing comprehensiveness with empirical depth. As mentioned before, the academic studies

---

[5] For example, see Daryl G. Press, "The Myth of Air Power in the Persian Gulf War and the Future of Warfare," *International Security*, Vol. 26, No. 2, Fall 2001; Stephen Biddle, "Afghanistan and the Future of Warfare," *Foreign Affairs*, Vol. 82, No. 2, March–April 2003; Stephen Biddle and Jeffrey A. Friedman, *The 2006 Lebanon Campaign and the Future of Warfare: Implications for Army and Defense Policy*, Carlisle, Pa.: Strategic Studies Institute, 2008.

[6] Ministry of Defence UK, 2014, p. 2; Modernisation and Strategic Planning Division, Australian Army Headquarters, 2014, p. 8–9; Department of the Air Force, 2015, pp. 3–4; U.S. Marine Corps, Futures Directorate, 2015, pp. 12, 52–62; Joint Chiefs of Staff, 2016, pp. 10–12; National Intelligence Council, 2017, pp. 8–10.

[7] Ministry of Defence UK, 2014, pp. 3, 8–9; Modernisation and Strategic Planning Division, Australian Army Headquarters, 2014, p. 9; Department of the Air Force, 2015, pp. 5–6; U.S. Marine Corps, Futures Directorate, 2015, pp. 27–34; Joint Chiefs of Staff, 2016, p. 11; National Intelligence Council, 2017, pp. 21–25.

[8] Ministry of Defence UK, 2014, pp. 1–2; Modernisation and Strategic Planning Division, Australian Army Headquarters, 2014, p. 11; Department of the Air Force, 2015, p. 7; U.S. Marine Corps, Futures Directorate, 2015, pp. 29–46; Joint Chiefs of Staff, 2016, pp. 13, 16; National Intelligence Council, 2017, pp. 13, 17.

tend to be detailed but limited in scope. By contrast, the military studies tend to be broader in scope but less empirically rich.

Second (and perhaps more fundamentally), these studies identify future global trends, but thinking through the specific implications of these trends for conflict can prove challenging. Even the National Intelligence Council's *Global Trends: Paradox of Progress* (which is probably the best of its kind at tackling a wide-ranging topic with a detailed, evidence-based approach) speaks to a broader audience outside the national security sphere—and, consequently, discussion of the future of warfare is relegated to a few pages at the end of the report.[9]

Our project—sponsored by the USAF—tries to chart a middle course. Similar to the other government-produced studies, this series of reports on the future of warfare adopts a comprehensive scope and focuses on many of the same political, military, economic, geographic, legal, and informational trends. By dividing these trends into separate thematic reports, however, we attempt to preserve the analytical rigor of academic studies while maintaining the brevity necessary to speak to a policy audience. In sum, our project is not necessarily methodologically distinct from its predecessors (in the sense it looks at a similar set of broad trends), but it does aim to strike a unique balance between breadth and depth with a focus on warfare.

Definitions and Methodology

Before delving into the substance of these trends, we must first start with two definitional questions: What is warfare, and what is the future? The former proves surprisingly challenging to define. *Warfare*, after all, can refer to any sort of conflict—terrorism, gray-zone attempts to undermine an adversary, large-scale conventional operations, or full-on nuclear exchanges. The definitional problem is particularly acute because many typical scholarly definitions are either infeasible or irrelevant. Traditionally, political science imposes a set, if arbitrary, standard to define wars—such as requiring 1,000 battle

[9] National Intelligence Council, 2017, pp. 215–221.

deaths per year—but casualty figures in future conflicts are hypothetical at best, making this bar of little value.[10]

Consequently, we adopt a Clausewitzian definition of *warfare* as "an act of force to compel our enemy to do our will."[11] It requires that war be a deliberate, coercive act applied against an adversary. It does not require that any such act inflict a set number of deaths or any physical damage whatsoever (particularly in our discussions in the next chapter of cyber war, information operations, and lawfare). To focus this study and maximize its value, we look at the future of warfare through a U.S. lens; specifically, how the U.S. military (specifically the USAF) might employ force in the future and how our adversaries might employ force against us in return.

Defining the term *future* proves easier, if somewhat arbitrary. Although different studies employ a variety of time frames, we default to a middle estimate of 2030 unless specifically noted otherwise.[12] In using this time line, we tried to strike a balance between accuracy and utility and to project out far enough to influence USAF capabilities development (given that major weapon systems can take years, even decades, to develop and field) but not so far into the future that accurate predictions become impossible.

The necessarily broad definitions of *future* and *war* make dividing the topic into discrete, analyzable elements a challenge. This series of reports adopts a thematic approach delving into how a variety of variables—geopolitical, military, economic, environmental, geographic, legal, and informational (the last two defined as *restraint*)—will shape the conduct of warfare and, specifically, the USAF's role in it. In each area, we identified about a half dozen of the top trends. The methods

---

[10] This is the standard for war, applied in one of the most common political science data sets. Meredith Reid Sarkees, "The COW Typology of War: Defining and Categorizing Wars (Version 4 of the Data)," paper, undated, as hosted on Meredith Reid Sarkees and Frank Wayman, *Resort to War: 1816–2007*, Washington, D.C.: CQ Press, 2010.

[11] Carl von Clausewitz, *On War*, ed. and trans. Michael Howard and Peter Paret, Princeton, N.J.: Princeton University Press, 1984, p. 75.

[12] For example, the National Intelligence Council's most recent attempt at long-term predictions uses both an 18-year (2035) and a five-year event horizon. See National Intelligence Council, 2017.

for selecting the specific trends in each area are detailed in the individual reports, but the determination generally was based on a review of other scholarly work, such as those studies mentioned earlier, an analysis of different data sets, client interest, extensive field research, and professional judgment.

Over the course of the project, we interviewed more than 120 different government, military, academic, and policy experts from more than 50 different institutions in Belgium, China, Germany, Israel, Japan, Jordan, Poland, the United Arab Emirates, and the United Kingdom for their perspectives on regional and global trends that might shape the future of conflict between now and 2030. As in any study of this scope, there were many more potential trends that could have been explored in depth but were not for reasons of time and resources. Nonetheless, each of these trends will shape the future of conflict in profound ways in the years to come, as we shall see in the next chapter.

CHAPTER FOUR

Depicting the Key Trends

The other volumes in this series document each trend and describe exactly how it will shape conflict. In this report, we summarize some of the basic findings in the narrative and tables.

The backdrop for our examination of conflict in 2030 is the geopolitical trends shaping the strategic environment (Table 4.1), starting with those within the world's remaining superpower, the United States. The American public is becoming increasingly polarized on a variety of issues—including foreign and defense policy—producing political gridlock in the United States (trend 1 in Table. 4.1).[1] This gridlock will limit U.S. ability to do the tasks necessary to act effectively as a global superpower, from resourcing the defense budget to responding to international crises in a coherent and unified manner. Just as troubling, politicians might increasingly look for military solutions because the military is one of the few government institutions that Americans trust.[2] These political ills show no sign of abating and could even increase in years to come.

The absence of U.S. leadership on the international stage would not be as serious if it were not for the other five trends. First, China is rising (trend 2 in Table 4.1). China's president Xi Jinping recently waived term limits and promised "the great rejuvenation of China"—restoring China to what it perceives as its rightful place on the world

[1] See Pew Research Center, "Political Polarization in the American Public," June 12, 2014.

[2] Gallup "Confidence in Institutions," 2017.

Table 4.1
Geopolitical Trends

Trend	Who Will Fight	How the United States Will Fight	Where the United States Will Fight	When the United States Will Fight	Why the United States Will Fight
1. U.S. polarization and retrenchment		Airpower and special operation forces rather than conventional ground forces, if the United States fights at all			Vacuum in U.S. leadership
2. China's rise	China and its immediate neighbors		Taiwan, South China Sea, Senkakus	If China's economy slows; potentially as Xi's tenure comes to a close	Domestic pressure; expanding strategic periphery
3. Asia's reassessment	New U.S. partners and allies (e.g., Vietnam/India); less others (e.g., the Philippines)	More maritime conflicts (air-sea cooperation)			Nationalism; fear of rising China
4. A revanchist Russia	Russia and its neighbors		Russia's near abroad (but with second-order effects for Asia and the Middle East)		Combination of Russian insecurity and desire for a greater sphere of influence
5. Upheaval in Europe	Varies based on country and type of conflict		Eastern Europe (in response to Russian aggression)		Counterterrorism; response to Russian aggression
6. Turmoil in Islamic world		Sustained low-level conflict/counterterrorism	Middle East, North Africa, Central Asia	Now ongoing	Counterterrorism

stage and reversing its "century of humiliation."[3] Second, as China rises, other states—particularly in Asia—are reacting to their larger, more powerful neighbor's growing ambitions by rethinking whether to get on the bandwagon with or to balance against China's rise (trend 3 in Table 4.1). Third, Russia, though arguably a declining power, also is growing more aggressive, intervening in Moldova, Georgia, Ukraine, and Syria and reaffirming its position as a great power (trend 4 in Table 4.1). Fourth, the European Union is becoming more fractured, less interested in expeditionary operations, and increasingly inward-looking, mired as it is in an immigration crisis, the growth of right-wing populism, and the lingering effects of the Euro crisis (trend 5 in Table 4.1).[4] Finally, despite a decades-long international counterterrorism campaign, the Middle East remains afflicted with Islamic jihadist terrorism, systemic poor governance, economic issues, and growing tensions between Iran and Saudi Arabia and between Iran and Israel that are already shaping conflicts in Syria, Yemen, and beyond (trend 6 in Table 4.1). None of these problems appear likely to be resolved any time soon and will likely shape the contours of conflict in the years to come.

The military trends mirror the geopolitical ones in many ways (Table 4.2). First, U.S. conventional overmatch is declining. Despite "reemergence of long-term, strategic competition," the U.S. military will likely remain a fraction of the size it was during the Cold War, which was the last period of long-term, strategic competition, and it will lack the technological superiority it enjoyed during the immediate aftermath of the Persian Gulf War.[5] Second, the Chinese and (to a lesser extent) Russian militaries are becoming increasingly capable,

[3] Graham Allison, "What Xi Jinping Wants," *The Atlantic*, May 31, 2017; interviews with multiple Chinese think tank analysts, Beijing, China, June 11–15, 2018.

[4] Interviews with government officials, international organization officials, and think tank officials in London, Brussels, and Berlin, April 16–20, 2018.

[5] DoD, *Summary of the 2018 National Defense Strategy of the United States of America: Sharpening the American Military's Competitive Edge*, Washington, D.C., 2018, p. 2.

Table 4.2
Military Trends

Trend	Who Will Fight	How the United States Will Fight	Where the United States Will Fight	Why the United States Will Fight
1. Decreasing U.S. conventional force size		Multidomain under nuclear shadow with some amount of artificial intelligence (AI)		Regional aggressor calculates that the United States lacks capacity to respond effectively in a given theater because of its other global commitments
2. Increasing near-peer conventional modernization and professionalization	China/Russia vs. United States and select allies or partners	Multidomain under nuclear shadow with some amount of AI	East China Sea, Taiwan, South China Sea, Baltics, or elsewhere on peripheries	China or Russia calculates it can deny the United States sufficient access to defeat effort to change territorial status quo
3. Selectively capable second-tier powers	Iran/North Korea vs. United States, allies and partners	Neutralize selective capabilities then destroy large but less-sophisticated forces	Middle East or Korean Peninsula	Iranian machinations/North Korean provocations lead to war
4. Adversary use of gray-zone tactics	Quasi-military or covert state forces, nonstate actors	Subconventional or hybrid, potentially escalating to conventional	In disputed territories and areas where state control is weak	States victimized by covert or proxy forces will need support
5. Weakening of the state's monopoly on violence	Heavily armed individuals and groups	Subconventional or hybrid	Areas of failed or weak state control—Africa, Middle East, South Asia	States unable to restrain heavily armed individuals and groups will need support
6. AI as a class of disruptive technologies	Highly advanced states	Multidomain under nuclear shadow with autonomous weapons		Regional aggressor believes its AI capabilities are sufficient to change the status quo

as both continue to modernize and professionalize.[6] In China's case, especially, these military improvements likely will continue, closing the qualitative gap between the People's Liberation Army and the U.S. military.[7] Third, second-tier powers—such as Iran and North Korea—cannot militarily match the United States and are instead increasingly turning to asymmetric capabilities—such as cyber, missiles, and weapons of mass destruction—to counter conventional U.S. superiority.[8] Fourth, and almost as important as the changes in their capabilities, U.S. adversaries are changing their tactics to operate in the gray zone, using coercion to achieve their national objectives below the threshold of war.[9] Part of the success of gray-zone tactics stems from a fifth trend: the weakening of state monopolies on violence. Thanks to changes in military and communications technology, nonstate actors—or in the case of gray-zone conflicts, proxy forces—can destabilize states with increasing ease.[10] Finally, new military technology—most notably AI—is on the horizon that could upset today's military balances. Taken together, these trends point to the fact that, as the *National Defense Strategy* argues, "competitive military advantage has been eroding" and, if unaddressed, will allow U.S. adversaries to exploit these weaknesses to their own advantages.[11]

---

6 See Eric Heginbotham, Michael Nixon, Forrest E. Morgan, Jacob L. Heim, Jeff Hagen, Sheng Li, Jeffrey Engstrom, Martin C. Libicki, Paul DeLuca, David A. Shlapak, David R. Frelinger, Burgess Laird, Kyle Brady, and Lyle J. Morris, *The U.S.-China Military Scorecard: Forces, Geography, and the Evolving Balance of Power, 1996–2017*, Santa Monica, Calif.: RAND Corporation, RR-392-AF, 2015; Scott Boston and Dara Massicot, *The Russian Way of Warfare: A Primer*, Santa Monica, Calif.: RAND Corporation, PE-231-A, 2017.

7 See Heginbotham et al., 2015.

8 See International Institute for Strategic Studies, *The Military Balance*, Vol. 118, No. 1, 2018, pp. 221, 275–277, 333–337.

9 For an analysis, see Michael J. Mazarr, *Mastering the Gray Zone: Understanding a Changing Era of Conflict*, Carlisle Barracks, Pa.: U.S. Army War College Press, 2015.

10 J. J. Messner, Nate Haken, Patricia Taft, Ignatius Onyekwere, Hannah Blyth, Charles Fiertz, Christina Murphy, Amanda Quinn, and McKenzie Horwitz, *2018 Fragile States Index*, Washington, D.C.: Fund for Peace, 2018.

11 DoD, 2018, p. 1.

In addition to the overall military balance, developments in the space and nuclear realms deserve special mention, especially given USAF equities in both areas (Table 4.3). The ability to use space-based assets for intelligence, communication, and navigation has long been one of the cornerstones of the U.S. military's advantage, but future U.S. dominance in space could be subject to two countervailing trends. First, space is becoming an increasingly contested environment as China and Russia improve their abilities to disable and destroy satellites (trend 1 in Table 4.3).[12] Second, the commercial exploitation of space has exploded in recent years and the trend is likely to continue through 2030 (trend 2 in Table 4.3).[13] As more commercial entities launch microsatellites for imagery and communications purposes, the overall U.S. space infrastructure could grow more resilient—provided the United States can leverage these commercial investments.

Nuclear trends present a cleaner, if less rosy, picture of the future. Nuclear proliferation is once again an international concern (trend 3 in Table 4.3). Several second-tier states—most notably Iran and North Korea—have pushed to develop nuclear weapons and despite concerted international diplomatic efforts to prevent nuclear proliferation (in the former's case) or roll it back (in the latter's), it remains unclear whether either effort will be successful.[14] Should these efforts fail, Iranian and North Korean nuclear proliferation might spur further regional nuclear proliferation, preventative war, and possibly even limited nuclear war, especially given the fraught relationships that both states have with their neighbors. At the same time, nuclear arms control regimes appear to be eroding, increasing the chances that Russia—and, to a lesser extent, China—might employ tactical nuclear weapons in the future (trend 4 in Table 4.3).[15]

---

[12] Daniel R. Coats, *Statement for the Record: Worldwide Threat Assessment of the U.S. Intelligence Community*, Washington, D.C.: Office of the Director of National Intelligence, February 13, 2018, p. 13.

[13] For market trends, see Bryce Space and Technology, "State of the Satellite Industry Report," Alexandria, Va., for the Satellite Industry Association, June 2017.

[14] Coats, 2018, pp. 7–8.

[15] Coats, 2018, p. 7.

Table 4.3
Space and Nuclear Trends

Trend	Who Will Fight	How the United States Will Fight	Where the United States Will Fight	Why the United States Will Fight
1. Space as an increasingly contested environment	China and Russia	Terrestrial- and space-based reversible- and nonreversible-effects attacks on space capabilities	In orbits where satellites operate and at terrestrial nodes of space systems' infrastructure	Interdiction of enemy space force enhancement capabilities and defense of U.S. capabilities
2. Proliferation of commercial space capabilities	China and Russia	Terrestrial-based attacks on space capabilities	At terrestrial nodes of space systems' infrastructure	Interdiction of enemy space force enhancement capabilities and defense of U.S. capabilities
3. Resumption of nuclear proliferation	North Korea, Iran, possibly others	Preventive or preemptive conventional strikes on enemy nuclear capabilities; conventional or nuclear counterstrikes	Northeast, South, or Southwest Asia	Escalation of conventional crises or conflicts
4. Erosion of norms/treaties constraining tactical nuclear weapons use	China and Russia	Preventive or preemptive conventional strikes on enemy tactical nuclear capabilities; conventional or nuclear counterstrikes	Europe or the Western Pacific	Escalation of conventional crises or conflicts

As the world becomes progressively digitized, cyber operations will play an increasingly vital role in warfare, particularly in three areas (Table 4.4). First, control of the cyber domain will become increasingly central to domestic stability. The most extreme example here is probably China, which tightly monitors the content that its citizens can access and uses cybersurveillance for behavior control, but all states are concerned about preventing the use of the cyber domain as a tool for foreign subversion.[16] Second, as more data are digitized and held in the cloud, the cyber domain will become the primary target of espionage efforts.[17] Finally, cyber sabotage could play an increasingly important role in warfare in 2030. In 2007, the U.S. intelligence community assessed that only a handful of countries possessed offensive cyber capabilities; in 2017, the number had grown to more than 30.[18] At the same time, much of the United States' critical infrastructure lies outside of the direct control of DoD—and of the U.S. government—and thus poses a comparatively easy target for adversaries to attack.[19]

All military capabilities, however, matter only to the extent that actors decide to employ them, which leads us to the question of the future of restraint in warfare (Table 4.5). A host of factors—such as international law, public opinion, media coverage, technological capabilities, partner preferences, and operational imperatives—shape the amount of restraint that combatants exercise in conflict, and many of these variables will increasingly weigh on how the United States—and its mostly liberal democratic allies—will fight wars in 2030. First, as smartphones and social media saturate the developing world, militaries will find themselves harder pressed to control both what images

---

[16] For example, see Elizabeth C. Economy, "The Great Firewall of China: Xi Jinping's Internet Shutdown," *The Guardian*, June 29, 2018; Coats, 2018, p. 11.

[17] Coats, 2018, p. 6; National Counterintelligence and Security Center, *Foreign Economic Espionage in Cyberspace*, Washington, D.C.: Office of the Director of National Intelligence, 2018.

[18] Coats, 2018, p. 5.

[19] See, for example, the President's National Infrastructure Advisory Council, *Securing Cyber Assets: Addressing Urgent Cyber Threats to Critical Infrastructure*, Washington, D.C., Department of Homeland Security, August 2017.

Table 4.4
Cyber Trends

Trend	Who Will Fight	How the United States Will Fight	Where the United States Will Fight	Why the United States Will Fight
1. Information control	Russia, China, Iran, North Korea and nonstate actors	Information operations to counter adversary narrative	Cyberspace	Prevent propaganda from influencing the U.S. public and causing domestic discord
2. Cyber espionage	China, Russia and nonstate actors	Strengthen cyber defenses; continue to develop better cyber intrusion detection methods	Cyberspace	Protect U.S. national security, intellectual property, research and development
3. Cyber sabotage	Russia, Iran, North Korea and nonstate actors	Build resilient, redundant networks	Cyberspace	Protect critical infrastructure and communication networks, prevent data destruction

Table 4.5
Restraint Trends

Trend	Who Will Fight	How the United States Will Fight	Where the United States Will Fight	When the United States Will Fight	Other Implications
1. Widespread distribution of imagery of military operations	Some greater deterrence of liberal and democratic states; others less affected	Greater importance of precision-guided munitions; micromunitions; intelligence, surveillance and reconnaissance (ISR); public communication	More likely in urban areas, despite U.S. desire to avoid	Potentially quicker adversary counterattacks	
2. Increasing public concern for civilian casualties	Greater deterrence of liberal and democratic states; others less affected. Potentially lower utility from U.S. partners	Greater importance of precision-guided munitions, micromunitions, ISR, public communication	More likely in urban areas, despite U.S. desire to avoid		
3. The spread of lawfare	Emboldened nonstate actors and autocracies; liberal-democratic states more deterred	Greater importance of precision-guided munitions, micromunitions, ISR, public communication	More likely in urban areas, despite U.S. desire to avoid		
4. Increasing power of false accusations		Transparent, reliable public affairs vital			Increased risk of mission failure because of lack of public support

the public sees and the narrative surrounding operations (trend 1 in Table 4.5).[20] This is particularly relevant because of a second reinforcing trend: Domestic opinion in liberal democracies is increasingly sensitive to civilian casualties, especially in perceived wars of choice (trend 2 in Table 4.5).[21] By contrast, the mostly authoritarian adversaries of the United States might not feel similarly constrained by their publics, by international opinion, or international law.[22]

U.S. adversaries are also increasingly adept at manipulating international law to capitalize on U.S. and allied restraint. Known as *lawfare*, or "the strategy of using (or misusing) law as a substitute for traditional military means to achieve an operational objective,"[23] adversaries (such as Hamas in Gaza, China in the South China Sea, and Russia in Ukraine) have relied on this strategy to confound U.S. and allied responses and will likely do so increasingly in the future (trend 3 in Table 4.5).[24] At the same time, the media in the United States have likely become more susceptible to disinformation because of the increasing role of social media, increasing distribution of opinions over facts

---

[20] Statista, "Number of Social Network Users Worldwide from 2010 to 2021 (in billions)," webpage, undated; Jefferies & Co., Global Technology, Media, and Telecom Team, "Mobility 2020: How an Increasingly Mobile World Will Transform TMT Business Models over the Coming Decade," New York, September 2011, p. 155; Cory McNair, *Worldwide Social Network Users: eMarketer's Estimates and Forecast for 2016–2021*, New York: eMarketer, July 12, 2017.

[21] Eric V. Larson and Bogdan Savych, *Misfortunes of War: Press and Public Reactions to Civilian Deaths in Wartime*, Santa Monica, Calif.: RAND Corporation, MG-441-AF, 2007, p. xviii; Stephen Watts, "Air War and Restraint: The Role of Public Opinion and Democracy," in Matthew Evangelista, Harald Muller, and Niklas Schorning, eds., *Democracy and Security: Preferences, Norms, and Policy-Making*, New York: Routledge, Taylor & Francis Group, 2008.

[22] As one interviewee stated, "We believe China is a great power and great powers should not be bound by international law." Interview with Chinese academic, Beijing, June 12, 2018.

[23] Charles J. Dunlap, Jr., "Lawfare: A Decisive Element of 21st-Century Conflicts? " *Joint Forces Quarterly*, No. 54, Third Quarter 2009, p. 35.

[24] James Kraska, *How China Exploits a Loophole in International Law in Pursuit of Hegemony in East Asia*, Philadelphia, Pa.: Foreign Policy Research Institute, January 22, 2015; Mazarr, 2015; Raphael S. Cohen, David E. Johnson, David E. Thaler, Brenna Allen, Elizabeth M. Bartles, James Cahill, and Shira Efron, *From Cast Lead to Protective Edge: Lessons from Israel's Wars in Gaza*, Santa Monica, Calif.: RAND Corporation, RR-1888-A, 2017.

in traditional media outlets, declining levels of trust in the government, and the increasing influence of explicitly partisan news sources. These developments will give adversaries more opportunities to spread disinformation and potentially to deter U.S. or allied military action (trend 4 in Table 4.5).[25] Taken together, the United States could confront a widening "restraint gap" between how the United States and its allies will use force in conflicts and how its adversaries will—particularly in wars waged on the lower ends of the conflict spectrum.

Similar to restraint, global economic trends play a significant, if indirect, role in shaping why and how wars could be fought in 2030 (Table 4.6). Three global economic trends increase the chances of future conflict. First, although analyses differ, greater trade is generally correlated with less war.[26] Notably, protectionism is on the rise, although trade remains far freer today than it has been throughout most of the post-WWII era (trend 1 in Table 4.6). Even before the recent rounds of trade tariffs between the United States and China, governments around the world had carried out more than 15,000 trade-related interventions between November 2008 and early 2018, most of them restraints.[27] Second, China's economic ambitions are expanding (trend 2 in Table 4.6). Its Belt and Road Initiative extends across Eurasia to traditional U.S. allies (such as the United Kingdom, France, and Germany), and as Chinese economic interests grow, so will Chinese security interests.[28] Finally, control of natural resources and access to them have long been considered a potential cause of war. The future global econ-

---

[25] Jennifer Kavanagh and Michael D. Rich, *Truth Decay: An Initial Exploration of the Diminishing Role of Facts and Analysis in American Public Life*, Santa Monica, Calif.: RAND Corporation, RR-2314-RC, 2018, pp. 95–121.

[26] Edward D. Mansfield, *Power, Trade, and War*, Princeton, N.J.: Princeton University Press, 1994.

[27] Simon J. Evenett and Johannes Fritz, *Going Spare: Steel, Excess Capacity, and Protectionism*, 22nd Global Trade Alert Report, London: CEPR Press, 2018.

[28] In a May 2018 publication, China counted 71 countries in addition to itself as being part of Belt and Road (Cheng Xiaobo, ed. [程晓波主编], *Big Data Report on Trade Cooperation under the Belt and Road Initiative* [一带一路贸易合作大数据报告], "Belt and Road" Big Data Center of the State Information Center [国家信息中心"一带一路"大数据中心] and SINOIMEX [大连瀚闻资讯有限公司], May 2018, p. ix.

Table 4.6
Global Economic Trends

Trend	Who Will Fight	How the United States Will Fight	Where the United States Will Fight	When the United States Will Fight	Why the United States Will Fight
1. Increasing pressure on the global trading system	China, United States		Asia, with possible worldwide spillovers	2020s	Fear of economic damage from closing markets
2. The rise of China	China, United States, Japan		Asia, with possible worldwide spillovers		Miscalculation; Chinese ambitions or opportunism or a China in crisis will go to war to spur cohesion and nationalism
3. The search for new resources	Smaller countries	Interventions in small wars and civil wars	Developing countries		To protect a partner under attack or to forestall monopolization of a resource
4. Relatively declining U.S. and allied economic might	Raises risk of challenge by Russia, China, Iran			Likely later 2020s	Miscalculation or opportunistic action to take advantage of perceived weakness
5. The shrinking defense industrial base		Readiness and resilience negatively affected; war duration might lengthen, temptation to use more-powerful weapons or deliver knockout blow might increase	Uncertain	Could lessen the frequency of war	
6. Decreasing power of sanctions	Raises risk of U.S. military action			If sanctions substitute for war, sooner than otherwise	

omy will require additional resources, such as energy, minerals, and resources for new technologies and industries, and this could increase the chances of conflict, especially if states cannot trade freely for these resources on the open market (trend 3 in Table 4.6).[29]

At the same time, global economic trends will also shape how wars are fought. On an absolute level, the United States and its allies will likely collectively remain the dominant force in the global economy in 2030.[30] Nonetheless, as China rises, the United States and its allies will rise more slowly and therefore make up a smaller share of global gross domestic product (GDP) (trend 4 in Table 4.6). As a result, as we look to 2030, the United States will be less able to rely on the overwhelming economic dominance it enjoyed in the latter half of the 20th century to give it a quantitative or even qualitative military advantage. This finding will be exacerbated by another trend—the consolidation and decline in both U.S. and allied defense industrial bases (trend 5 in Table 4.6).[31] With fewer warm production lines and fewer types of aircraft and other major equipment being produced, the United States and its allies will face fewer choices in 2030 for major weapon systems and a diminished capacity to ramp up production that might be needed for a major conflict. Finally, the power of the United States to use its preferred means of coercion, economic sanctions, might decline if other major economies develop alternative means of international payments in reaction to overuse, if coordination among allies becomes more difficult, and if China makes its financial sector far more open than it is now (trend 6 in Table 4.6). If sanctions become progressively more ineffective, the United States might need to resort to more-kinetic forms of coercion.

---

[29] Matthew O. Jackson and Massimo Morelli, "The Reasons for Wars: An Updated Survey," in Christopher J. Coyne and Rachel L. Mathers, eds., *The Handbook on the Political Economy of War*, Cheltenham, UK, and Northampton, Mass.: Edward Elgar, 2011.

[30] Projections based on data from World Bank, World Development Indicators, June 28, 2018.

[31] Marjorie Censer, "Defense Companies Brace for a Different Kind of Consolidation This Time Around," *Washington Post*, January 12, 2014.

Finally, the future of warfare will also be shaped by several environmental trends (Table 4.7). Although the impact of climate change will be mostly felt in the far future of 2050 and beyond, some impacts could begin to manifest themselves sooner.[32] Global air surface temperatures will likely be 1 degree Fahrenheit warmer in 2030 than they were in the latter decades of the 20th century, affecting health, reducing economic productivity, and contributing to a host of operational problems for basing aircraft in already hot parts of the globe, such as the Persian Gulf (trend 1 in Table 4.7).[33] Hotter temperatures also can cause a series of equally problematic second-order effects. They can exacerbate potable water shortages, including in places already prone to instability and substate violence—particularly in the Middle East, sub-Saharan Africa, and parts of Asia (trend 2 in Table 4.7).[34] Melting polar ice will make the Arctic more navigable and likely increase the chances of spillover conflict in the area among rival great powers—the United States, Russia, and China (trend 3 in Table 4.7).[35] At the same time, rising sea levels will cause humanitarian challenges and shift the geography in geopolitically sensitive regions, such as the South China Sea, affecting Chinese sovereignty claims (trend 4 in Table 4.7).[36] Extreme weather events will not only increase the demand for disaster relief missions but also affect low-lying U.S. military bases, including

[32] R. K. Pachauri and L. A. Meyer, eds., *Climate Change 2014: Synthesis Report. Contribution of Working Groups I, II and III to the Fifth Assessment Report of the Intergovernmental Panel on Climate Change*, Geneva, Switzerland: Intergovernmental Panel on Climate Change, 2014.

[33] David Herring, "Climate Change: Global Temperature Projections" webpage, March 6, 2012.

[34] World Bank, *High and Dry: Climate Change, Water, and the Economy*, Washington, D.C., 2016.

[35] Stephanie Pezard, Abbie Tingstad, Kristin Van Abel, and Scott Stephenson. "Maintaining Arctic Cooperation with Russia: Planning for Regional Change in the Far North," RR-1731-RC, Santa Monica, Calif.: RAND Corporation, 2017.

[36] Wilson T. VornDick, "Thanks Climate Change: Sea-Level Rise Could End South China Sea Spat," *The Diplomat*, November 8, 2012; Wilson T. VornDick, "China's Island Building + Climate Change: Bad News," *RealClearDefense*, March 10, 2015; Steve Mollman, "It's Typhoon Season in the South China Sea—and China's Fake Islands Could Be Washed Away," *Quartz*, August 1, 2016.

Table 4.7
Environmental Trends

Trend	Who Will Fight	How the United States Will Fight	Where the United States Will Fight	Why the United States Will Fight
1. Rising temperatures		Health risks to service members; maintenance challenges to air bases and aircraft	Arid, warm regions where ethnic socioeconomic tensions already exist (Middle East and North Africa [MENA], East Africa)	Reduced economic productivity could weaken governments and exacerbate ethnic tensions
2. Water scarcity		More humanitarian assistance disaster relief missions. Need for more cost-efficient water and energy uses	MENA, sub-Saharan Africa, and Central and South Asia	More water-related intrastate and interstate conflict and unrest
3. Opening of the Arctic	Russian and Chinese will increase presence	More Arctic ISR and reliable communications and mapping		Spillover from other conflicts likely started for reasons outside of Arctic issues
4. Sea level rise		Rising sea levels could affect USAF basing and training and increase demand for humanitarian assistance and disaster relief	Sea level rise could spark conflict in South China Sea and affect Chinese efforts to set maritime boundaries	Sea level rise acts as a threat multiplier. Displacement could strengthen violent nonstate actors
5. Extreme weather events		Flooding will likely affect basing and training; more humanitarian assistance, disaster relief, and counterterrorism missions	Bases in Asia Pacific region are most vulnerable	Could increase spread of disease and migration instigating spikes in violence
6. Water scarcity		USAF could face more demand for humanitarian assistance and disaster relief missions; need to adopt more cost-efficient water and energy uses	Countries unable to mitigate water scarcity in the MENA region, Sub-Saharan Africa and Central and South Asia.	Could hinder food security and undermine livelihood in agriculture dependent areas, leading to more water-related intrastate and interstate conflict and unrest
7. Urbanization and megacities	Terrorist groups, gangs, warlords	Decreased advantage from air superiority; more nonlethal, high-precision weapons	Megacities in developing countries	Failed governance and lawlessness, which might require U.S. intervention

those in strategic locations that are already at risk of flooding, such as the Marshall Islands, Guam, and Diego Garcia (trend 5 in Table 4.7).[37]

Geography will shape conflict in other ways. Global population is becoming more urbanized (trend 6 in Table 4.7). For the first time in 2008, more than half of the world's population lived in cities, and the number is growing.[38] By 2030, the number of megacities—those with 10 million or more inhabitants—will expand from 31 to 41.[39] As populations become more urbanized, particularly in the developing world, states will be harder pressed to maintain law and order, and militaries in general—and airpower in particular—will face a more difficult challenge of discriminating between military and civilian targets.[40]

Ultimately, the subsequent volumes in these series explore each of the aforementioned future trends in detail. We outline the evolution of each trend's connection to conflict in general, how the variable in question has shaped conflict in the past, how it will do so in the future, and what the implications of the trend are for the USAF in particular and the joint force at large. Each of these trends, however, provides only a partial glimpse into the future. The next chapter looks at these trends collectively and tries to paint a holistic picture of the future of warfare.

---

[37] Heather Messera, Ronald Keys, John Castellaw, Robert Parker, Ann C. Phillips, Jonathan White, and Gerald Galloway, *Military Expert Panel Report: Sea Level Rise and the U.S. Military's Mission*, 2nd ed., Washington, D.C.: Center for Climate and Security, 2018.

[38] James Canton, "The Extreme Future of Megacities," *Significance*, Vol. 8, No. 2, 2011, p. 53.

[39] United Nations, Department of Economic and Social Affairs, Population Division, *The World's Cities in 2016—Data Booklet*, New York, 2016, p. 4.

[40] Some of these challenges could be offset by technology, especially in advanced countries. Advances in cameras, unmanned aerial vehicles, and AI will allow governments to maintain situational awareness over urban areas in new ways. By contrast, governments in the developing world that lack these tools will face greater challenges in controlling megacities.

CHAPTER FIVE

Predicting the Future of Warfare

Who will the United States fight against and which nations will be on which side? Where will these future conflicts be fought? What will future conflicts look like? How will they be fought? And why will the United States go to war? Not all the trends outlined in the previous chapter are relevant to every question, collectively, but they do provide pieces to the puzzle. In this chapter, we invert the columns and rows from the tables in the previous chapter, highlight which of the thematic trends answer each of these broad questions about the future of warfare, and try to knit them together into a cohesive narrative about the future of conflict in the coming years.

Who Fights? Fixed Adversaries; Allies in Flux

In the topsy-turvy world of foreign policy, the stated list of U.S. adversaries has remained remarkably constant. In 2016, Secretary of Defense Ash Carter testified that the United States confronts five principal adversaries—China, Russia, Iran, North Korea, and terrorist groups.[1] Although Carter's declaration was more explicit than those made in previous presidential administrations, a similar list of challenges can be found in strategic documents dating to the later years of the George

---

[1] Lisa Ferdinando, "Carter Outlines Security Challenges, Warns Against Sequestration," DoD News, Defense Media Activity, March 17, 2016.

W. Bush administration.[2] Despite a shift in political party and foreign policy orientation, the Donald Trump administration kept the same list of stated adversaries in both its *National Security Strategy* and its *National Defense Strategy*.[3] The continuity in the list of adversaries is all the more striking because the Barack Obama administration made the Russian "reset," a pivot to Asia (and away from wars in the Middle East), and, later, the Iran nuclear agreement the cornerstones of its foreign policy legacy. More recently, the Trump administration has tried to improve relations with Russia and North Korea. Indeed, despite multiple efforts by different presidential administrations over the past several decades to bring various adversaries into the proverbial fold, U.S. adversaries only rarely disappear as a strategic challenge.[4] Looking forward, the trends outlined in the previous chapter suggest that all five actors mentioned in the *National Defense Strategy* will likely remain adversaries over the coming decade (Table 5.1) for a host of geopolitical, military, nuclear, economic, and environmental reasons.

Some of the reasons for this strategic continuity are structural. China and Russia view the existing international order as dominated by the United States and, at some level, contrary to their interests and security, so they want to change it.[5] Both countries prefer a more multipolar world with their own countries exerting more influ-

---

[2] See for example, for example, DoD, *Quadrennial Defense Review Report*, Washington, D.C., February 6, 2006, pp. 28–32.

[3] White House, *National Security Strategy of the United States of America*, Washington, D.C., December 2017, pp. 2–3; DoD, 2018, p. 4.

[4] Perhaps, the best examples of an adversary disappearing were Saddam Hussein's Iraq in 2003 and Moammar Qaddafi's Libya in 2011. In both cases, the leaders' tenures were ended through regime change.

[5] Andrew Radin and Clint Reach. *Russian Views of the International Order*, Santa Monica, Calif.: RAND Corporation, RR-1826-OSD, 2017; Michael J. Mazarr, Jonathan Blake, Abigail Casey, Tim McDonald, Stephanie Pezard, and Michael Spirtas, *Understanding the Emerging Era of International Competition: Theoretical and Historical Perspectives*, Santa Monica, Calif.: RAND Corporation, RR-2726-AF, 2018; interviews with retired Chinese general officers, Beijing, June 15, 2018.

Table 5.1
Fixed Adversaries

Category	Trend	China	Russia	Iran	North Korea	Terrorist Groups
Geopolitical trends	Rising China	√				
	A revanchist Russia		√			
	Terrorism, weak states, and proxy wars in Islamic world			√		√
Military trends	Increasing near-peer conventional modernization and professionalization	√	√			
	Selectively capable second-tier powers			√	√	
	Weakening of the state's monopoly on violence					√
Nuclear trends	Resumption of nuclear proliferation			√	√	
	Erosion of norms and treaties constraining tactical nuclear weapons use	√	√			
Global economic trends	Increasing pressure on the global trading system	√				
	The rise of China	√				
	The search for new resources	√	√			
Environmental trends	Opening of the Arctic	√	√			
	Urbanization and megacities					√

ence globally—and especially over their own regions.[6] Both countries' self-defined regional areas of influence butt up against those of the United States and its allies in key places, such as Taiwan, the East and South China Seas, Eastern Europe, and the Caucasus.[7] Consequently, although different U.S. presidential administrations might moderate their tones on both powers, Russia and China will likely remain U.S. adversaries on some level for at least the next decade or beyond.

As also noted in Table 5.1, there are also more indirect reasons to expect that China and Russia will continue to be U.S. competitors well into the next decade. On a military front, both China and Russia will be better positioned militarily to compete with the United States over the coming decade; consequently, they might be more willing to challenge international norms they find unfair or distasteful.

In the economic domain, China will need to acquire new sources of raw materials to drive its economic development and will seek these resources in different parts of the globe. As a result of this and other trade conflicts, economic rivalries will likely increase and feed the security competition between the great powers.

There are similar systemic reasons to believe that Iran and North Korea will persist as U.S. adversaries through 2030—absent state collapse or regime change, which are historically rare events. Similar to Russia and China, Iran and North Korea want to exert influence in their respective regions in ways that directly conflict with U.S. interests. Unlike Russia and China, however, neither Iran nor North Korea can militarily challenge the United States directly, although both have invested in asymmetric capabilities to do so. Finally, although both regimes have been the focus of U.S. diplomacy, both will likely remain U.S. adversaries. Despite President Obama's hope that the Joint Comprehensive Plan of Action regarding Iran's nuclear weapons program would "usher in a new era in U.S.-Iranian relations," even the deal's

---

[6] President of the Russian Federation, *Strategiya natsional'noi bezopasnosti Rossiiskoi Federatsii [National Security Strategy of the Russian Federation]*, Moscow, Decree No. 683, December 31, 2015; Graham Allison, *Destined for War: Can America and China Escape Thucydides's Trap?* New York; Houghton Mifflin Harcourt, 2017; Radin and Reach, 2017, p. ix; interviews with Chinese think tank officials, Beijing, June 12–15, 2018;.

[7] National Intelligence Council, 2017, pp. 35–36, 91, 125.

supporters acknowledged that Iran continued to conduct missile tests and provide support for its proxy forces in Yemen and Syria after the agreement was signed.[8] Although it is too early to say what, if anything, will come of the Trump administration's negotiations with North Korea, Chinese and Japanese security experts share a general skepticism regarding lasting progress in that arena.[9] Moreover, publicly released satellite imagery has indicated that, even after the summit, North Korea continues to expand its missile capabilities and one of its major nuclear research centers.[10]

Finally, despite the sustained U.S. counterterrorism effort since 2001, terrorist groups in general—and Islamic terrorist groups in particular—will continue to threaten U.S. interests through 2030, perhaps beyond. Indeed, the National Intelligence Council predicts, "The threat from terrorism will expand in the coming decades as the growing prominence of small groups and individuals use new technologies, ideas, and relationships to their advantage."[11] There are several reasons to believe the council's assessment. The Islamic State's so-called caliphate might be gone, but the ideology that drove it is not. Many predict the Sunni-Shi'a schism will intensify, which could drive Islamic terrorism in the Levant.[12] Finally, states in the developing world (particularly in the Middle East) will have to cope with protracted political, economic, and environmental challenges, and public discontent is likely to continue to fuel terrorism.[13] These same factors will weaken the governments in these states, which could

---

[8] Kenneth Katzman, *Iran: Politics, Human Rights, and U.S. Policy*, Washington, D.C.: Congressional Research Service, RL32048, May 21, 2018, pp. 21–22.

[9] Interviews with Chinese experts, Beijing, June 11–15, 2018; interviews with Japanese experts, Tokyo, June 18–19, 2018.

[10] Jonathan Cheng, "North Korea Expands Key Missile-Manufacturing Plant," *Wall Street Journal*, July 1, 2018.

[11] National Intelligence Council, 2017, p. x.

[12] National Intelligence Council, 2017, p. 41; interviews with Jordanian policy analysts and military officers, Amman, May 12–13, 2018.

[13] Interviews with Jordanian policy analysts and military officers, Amman, May 12–13, 2018; interviews with U.S. experts, Abu Dhabi, May 15, 2018.

make mounting an effective counterterrorism response and generally maintaining law and order progressively more challenging.

So, on the one hand, the list of U.S. adversaries will likely remain relatively constant from now until 2030. On the other hand, the United States could see significant changes among its partners and allies. For most of modern history, the United States has gone to war alongside its allies; if past is prologue, the United States will likely fight its next conflict with its allies and partners. From the trends outlined earlier, however, it is an open question exactly who those allies and partners will be (see Table 5.2).

In Asia, China's rise is forcing the countries of the region to reassess their alliances. Some—including historically nonaligned countries, such as India, and even previous adversaries, such as Vietnam—are opting for closer military relationships with the United States.[14] By contrast, other historical U.S. allies, most notably the Philippines, have chosen to distance themselves from Washington to pursue a cozier relationship with Beijing.[15] Indeed, in an October 2016 trip to China, Filipino President Rodrigo Duterte proclaimed that "I've realigned myself in your ideological flow," and supposedly offered a three-way alliance with China and Russia.[16]

There are serious headwinds to any potential realignment of U.S. alliances in Asia. For all the U.S. efforts to court India, it still retains enough of its nonaligned roots to make it suspicious of Western alliances.[17] By contrast, although Duterte might be interested in courting China, the generally pro-U.S. attitudes among the Filipino public, especially

---

[14] Patrick M. Cronin, Richard Fontaine, Zachary M. Hosford, Oriana Skylar Mastro, Ely Ratner, and Alexander Sullivan, *The Emerging Asia Power Web: The Rise of Bilateral Intra-Asian Security Ties*, Washington, D.C.: Center for a New American Security, 2013, p. 24; K. Alan Kronstadt and Shayerah Ilias Akhtar, *India-U.S. Relations: Issues for Congress*, Washington, D.C.: Congressional Research Service, R44876, June 19, 2017, pp. 1, 14, 17.

[15] Michael Auslin, "Duterte's Defiance," *Foreign Affairs*, November 2, 2016.

[16] Emily Rauhala, "Duterte Renounces U.S., Declares Philippines Will Embrace China," *Washington Post*, October 20, 2016.

[17] For example, although the United States and India published a joint strategic vision in 2015 that contained a thinly veiled warning to China about its actions in the South China Sea and the pair has also conducted joint military exercises, activities stopped short of a full-

Table 5.2
Allies in Flux

Category	Trend	Who Will Fight	Implications
Geopolitical trends	Rising China	China vs. its immediate neighbors	Potential for new alliances in Asia
	Growing tensions in Asia	Japan, India, Taiwan, Vietnam, and the Philippines (to a lesser extent) vs. China	Potential for new alliances in Asia
	A revanchist Russia	Potentially, countries in Russia's near abroad	Continuity in North Atlantic Treaty (NATO) allies that feel threatened by Russia
	Turmoil in Europe	Varies based on country and type of crisis, with Eastern Europe often showing the most will to oppose Russia	Potentially less contribution from traditional Western European allies
Military trends	Increasing near-peer conventional modernization and professionalization	China and/or Russia vs. United States and select allies or partners	Potential for new alliances in Asia among states that feel threatened by China; continuity in NATO allies that feel threatened by Russia
Restraint	Increasing public concern for civilian casualties	Greater deterrence of liberal and democratic states; others less affected. Potentially lower utility from U.S. partners	Potentially less contribution from traditional Western European allies
	The spread of lawfare	Emboldened nonstate actors and autocracies; liberal-democratic states more deterred	Potentially less contribution from traditional Western allies

the national security elite, likely limit how far this rapprochement can go—not to mention the Philippines' own conflicting territorial claims with China in the South China Sea.[18] Nonetheless, these changes indicate that, depending on the circumstances, the United States in 2030 could end up fighting in Asia with a very different coalition than it would have a decade ago.

In Europe, the United States confronts a somewhat different problem. Unlike in Asia, few, if any, European countries want to break with the United States and realign with Russia or any other U.S. adversary. To the contrary, even some of the governments that are most favorably disposed to Moscow still want to stay firmly rooted within the U.S. defense orbit.[19] The return of a revanchist Russia breathed new life into the NATO alliance—the cornerstone of the United States' strategic architecture. If anything, recent activities have renewed NATO's reason to exist.

That said, U.S. allies in Europe are undergoing changes in terms of will and capacity to exert force, particularly overseas. Despite a push to boost defense spending after Russia's invasion of Ukraine, many European countries—particularly some of the wealthiest ones—find themselves financially stretched, overextended, and struggling to resource their commitments of 2 percent of GDP to NATO.[20] Furthermore, Europe remains divided over a host of strategic questions, such as how to respond to Russia, and European nations are increasingly focused inward on internal problems, such as migration and terrorism.[21] Finally, many Western Europeans are particularly sensitive to

---

on military alliance. Office of the Press Secretary, White House, "U.S.-India Joint Strategic Vision for the Asia-Pacific and Indian Ocean Region," Washington, D.C., January 25, 2015.

[18] Richard Javad Heydarian, "Duterte's Dance with China," *Foreign Affairs*, April 26, 2016.

[19] For example, see the discussion of Slovakia and Hungary in Christopher S. Chivvis, Raphael S. Cohen, Bryan Frederick, Daniel S. Hamilton, F. Stephen Larrabee, and Bonny Lin, *NATO's Northeastern Flank: Emerging Opportunities for Engagement*, Santa Monica, Calif.: RAND Corporation, RR-1467-AF, 2017, pp. 44–66, 94–115.

[20] International Institute for Strategic Studies, 2018, pp. 70–72.

[21] Interviews with government officials, international organization officials, and think tank officials, London, Brussels, and Berlin, April 16–20, 2018.

lawfare, are demonstrating a growing sensitivity to casualties, and are less likely to support military intervention and the use of force—especially in the wake of the Iraq and Afghanistan wars. Consequently, Europe might demonstrate collectively less will to engage in expeditionary operations abroad, particularly when it comes to responding to potential Chinese aggression in Asia; Europe, as a whole, tends to downplay China as a military threat.[22]

The net result is that U.S. alliances in Europe will evolve, although possibly more in practice than on paper (and perhaps to a lesser degree than alliances in Asia). NATO might grow stronger as an entity in the coming decade, but its activities, ironically, likely will be directed more toward preserving security in Europe and away from expeditionary operations. Similarly, new member countries—such as Poland and other Eastern European countries that feel particularly threatened by Russia and thus most value U.S. military cooperation—might become increasingly willing to join U.S. alliances abroad while more-traditional Western European allies become more hesitant.[23]

Where Will the United States Fight? Most Likely Versus Most Dangerous

Despite the uncertainty of U.S. allies and partners, the relative stability of the list of U.S. adversaries offers some indication of where the next U.S. conflict might occur—at least regionally. Indeed, from the relevant trends outlined earlier, three major regions—Asia, Europe, and the Middle East—are all likely areas for the next war (Table 5.3). Unfortunately for military planners, going much beyond this level of fidelity proves difficult. Even a rough ordinal ranking of the theaters proves challenging. A list that sorts regions in terms of where the next conflict is most likely to occur greatly differs from a list that sorts

---

22 Interviews with government officials, international organization officials, and think tank officials, London, Brussels, and Berlin, April 16–20, 2018.

23 Interviews with government officials, international organization officials, and think tank officials, London, Brussels, and Berlin, April 16–20, 2018.

Table 5.3
Where the United States Will Fight

Category	Trend	Where the United States Will Fight	Asia	Europe	Middle East	Elsewhere
Geopolitical trends	Rising China	Taiwan, South China Sea, Senkakus	√			
	A revanchist Russia	Russia's near abroad (but second-order effects for Asia and Middle East)		√		
	Terrorism, weak states, and proxy wars in Islamic world	Middle East, North Africa, Central Asia	√		√	
Military trends	Increasing near-peer conventional modernization and professionalization	East China Sea, Taiwan, South China Sea, Baltics, or elsewhere on peripheries	√	√		
	Selectively capable second-tier powers	Middle East or Korean Peninsula			√	
	Weakening of the state's monopoly on violence	Areas of failed or weak state control—Africa, Middle East, South Asia	√		√	
Nuclear trends	Resumption of nuclear proliferation	Northeast, South, or Southwest Asia	√		√	
	Erosion of norms/treaties constraining tactical nuclear weapons use	Europe or the Western Pacific	√	√		

Table 5.3—Continued

Category	Trend	Where the United States Will Fight	Asia	Europe	Middle East	Elsewhere
Global economic trends	Increasing pressure on the global trading system	Asia, with possible worldwide spillovers	√			
	The rise of China	Asia, with possible worldwide spillovers	√			
	The search for new resources	Developing countries			√	√
Environmental trends	Rising temperatures	Arid warm regions where ethnic socioeconomic tensions already exist (MENA, East Africa)			√	√
	Sea level rise	Sea level rise could spark conflict in South China Sea and affect Chinese efforts to set maritime boundaries	√			
	Water scarcity	MENA, sub-Saharan Africa, and Central and South Asia	√		√	√

regions in terms of where conflict will likely be most dangerous for U.S. interests.

The 2018 *National Defense Strategy* prioritizes "deter[ring] aggression in three key regions—the Indo-Pacific, Europe, and Middle East."[24] Arguably, the ordering of the regions is no accident. With China, Russia, North Korea, and terrorist groups all militarily active in the Indo-Pacific region and with the first three being nuclear powers, a conflict in the Indo-Pacific region is likely the *most dangerous scenario* that the United States confronts today, and the danger will likely only increase in the future. As the *National Defense Strategy* predicts, China "will continue to pursue a military modernization program that seeks Indo-Pacific regional hegemony in the near-term and displacement of the United States to achieve global preeminence in the future."[25] Beyond the military balance questions, there are a host of unresolved territorial disputes (including some between nuclear armed rivals) in the East and South China Seas and along the India-China and India-Pakistan borders that could provoke confrontations, as well as a war on the Korean Peninsula that as of November 2018 still had not officially ended.

Asia is certainly preparing itself for conflict. Military spending across the region is up by 59 percent in real terms from 2008—one of the largest relative increases anywhere on the globe.[26] Although China ramped its defense expenditures by 110 percent over the past decade, it was by no means the only Asian power to boost defense spending.[27] Indeed, Cambodia, Bangladesh, and Indonesia increased their defense budgets by even larger percentages than China did, and Vietnam, the Philippines, India, Pakistan, and others all raised their defense budgets by 40 percent or more over the past decade.[28] Although countries spend on their militaries for a host of reasons, this trend on aggregate has two

---

[24] DoD, 2018, p. 6.

[25] DoD, 2018, p. 2.

[26] Nan Tian, Aude Fleurant, Alexandra Kuimova, Pieter D. Wezeman, and Siemon T. Wezeman, "Trends in World Military Expenditure, 2017," SIPRI Fact Sheet, May 2018, p. 5.

[27] Tian et al., 2018, p. 5.

[28] Tian et. al., 2018, p. 5.

major implications: First, Asian countries are sufficiently concerned about the prospects of future conflict to trade domestic spending for defense spending; second, if there is a war, it portends to be a bloody one, if for no other reason than the sheer amount of arms in the region.

Although the Indo-Pacific might be the most dangerous region for conflict, it might not be the most likely theater for overt use of U.S. military force, for two reasons. First, many disputes that would most likely trigger U.S. intervention—over the Senkakus (islands claimed by Japan and China), Taiwan, or the South China Sea—are maritime in nature, allowing the combatants to keep a potential conflict contained and below the threshold of war.[29] Many of these scenarios also would likely be interstate—rather than intrastate—conflicts between the United States and China, North Korea, or Russia, which would likely increase the chances of successful deterrence, especially because all these countries are armed with nuclear weapons. To be clear, this does not mean that the United States can categorically discount the possibility of such a conflict, nor does it say anything about the possibility of conflict between U.S. and adversarial forces below the threshold of conventional war. Rather, it could mean that the *most likely* shooting conflict for the United States might lie elsewhere.

Instead, of the three locations discussed, the Middle East arguably remains the most likely location for the United States to be involved in an overt, kinetic conflict from now until 2030. First, the United States has been fighting in the region since at least 2001 despite multiple presidential administrations' attempts to end U.S. military involvement. Second, the physical caliphate of the Islamic State might be gone, but the region is still in tatters after the Iraq War, Arab Spring, and the Syrian and Yemeni civil wars. Islamic jihadist terrorism also remains. The death toll of the Syrian conflict alone is so high that no definitive number exists (the United Nations places it at 400,000) and millions

[29] Political scientist John Mearsheimer refers to this phenomenon as "the stopping power of water." See John J. Mearsheimer, *The Tragedy of Great Power Politics*, New York: W. W. Norton, 2001, pp. 83–84.

more remain displaced.[30] Experts in neighboring Jordan expect it to be years, possibly even a decade or more, before Syria stabilizes.[31]

More crises might be on the horizon. Jordan is struggling under the economic weight of 1.5–2 million refugees, and there is near unanimous concern among Jordanian experts—inside and outside the military—that these stresses could fuel terrorism and instability in the future.[32] Iran's role in the Syrian conflict means Iranian forces are dangerously close to an Israeli redline, raising the chances of Israeli-Iranian conflict.[33] Similarly, Russia's increased presence in Syria might create another front in a broader U.S.-Russian competition. Finally, although many experts are supportive of Crown Prince Muhammad Bin Salman's attempt to reform Saudi Arabia, U.S. allies from Israel to Abu Dhabi worry about a potential backlash, and even a destabilization of the monarchy. Instability in Saudi Arabia could have second-order effects on the entire region.[34]

Beyond the geopolitics, however, there are environmental and economic reasons to expect continued instability in the Middle East through 2030. As global temperatures rise, the region's water scarcity problems will increase. Extreme heat already causes problems for civil aviation in the Gulf States—a key pillar of the economy—and rising temperatures will only add additional strain.[35] Lower prices for

---

[30] Megan Specia, "How Syria's Death Toll Is Lost in the Fog of War," *New York Times*, April 13, 2018.

[31] Interviews with a Jordanian academic and a Jordanian political analyst, Amman, May 12, 2018.

[32] Interviews with Jordanian academics, policy analysts, and military officers, Amman, May 12–13, 2018. Refugee figures come from an interview with a senior Jordanian army officer.

[33] For example, see the remarks of Israeli Minister of Defense Avigdor Lieberman at the 2018 Herzliya Conference after an Israeli strike on Iranian positions in Syria the previous day. Noa Shpigel, "Friction in North Not Over After Extensive Syria Strikes, Israeli Defense Minister Warns," *Haaretz*, May 11, 2018.

[34] Roundtable with Israel experts, Herzliya, Israel, May 9, 2018; interview with U.S. experts in Abu Dhabi, May 15, 2018. Similarly, see David A. Graham, "The Fragile Future of Reform in Saudi Arabia," *The Atlantic*, June 27, 2018.

[35] Interview with U.S. experts in Abu Dhabi, May 15, 2018.

fossil fuels have already hurt the Gulf economies and might eventually threaten to upset the socioeconomic structure of these states.[36] None of these trends are likely to cause conflict in and of themselves, but they are likely to exacerbate instability in the Middle East from now until 2030 when combined with broader geopolitical dynamics. Such instability could force a U.S. military response, albeit mostly in the context of counterterrorism or stability operations.

Finally, consideration must be given to where Europe stands relative to the other theaters. Although Europe faces an Islamic terrorism threat today, Russian military action remains the only foreseeable event that could prompt the United States to fight a war in Europe in 2030. Russia is a potentially dangerous foe, with its professionalizing military, modernizing nuclear arsenal, already-proven willingness to use force in Ukraine and Syria, and geographic advantage in striking NATO allies in Eastern Europe—particularly places, such as the Baltics, that are comparatively close to Russia but relatively difficult for the United States to readily defend. Furthermore, the ongoing conflict in Ukraine opens the possibility of escalation between the two powers into broader conflict.

In terms of danger, war in Europe likely ranks below war in the Indo-Pacific (if only because a war in the latter could involve both Russia and China) but above war in the Middle East. In terms of the likelihood of war, Europe ranks below the Middle East and possibly below the Indo-Pacific, although by how much is debatable. So far, Russia and the United States have avoided direct military confrontation over Ukraine and both seem content to keep the conflict limited. The power of interstate nuclear deterrence and the fact that Russia—unlike China—remains a power in decline could make large-scale war less likely in Europe than in the other theaters.

Ultimately, the mismatch between the most-likely and the most-dangerous places that the United States might fight wars means that U.S. defense strategists will face an ongoing conundrum in allocating resources: Do they prepare for the wars that the United States almost assuredly will fight? Or do they prepare for the wars that the United States hopes to avoid at all costs?

---

[36] Interview with U.S. experts in Abu Dhabi, May 15, 2018.

What Will Future Conflict Look Like? Four Archetypes of Conflicts

Perhaps, the best indicator for what future wars might look like is what capabilities U.S. adversaries are investing in today. After all, military modernization programs often take years to come to full fruition; in many cases, investment choices made today will continue to reverberate for years to come. Each future conflict will have its own unique flavor, but this analysis suggests that on aggregate, the United States will confront four basic archetypes of conflict (Table 5.4).

First, the United States will face a *counterterrorism fight*. As already mentioned, a host of geopolitical, economic, and environmental reasons make it likely that the current instability afflicting the Middle East and other parts of the world will continue for years—and with it, the ongoing threat of international terrorism. The United States will face what the National Intelligence Council predicts will be an increasingly "atomized jihadist militancy" that leverages advances in communication technology and the proliferation of conventional weaponry to wage a low-level, if decentralized, global terrorist campaign against the United States and its allies.[37] Consequently, the United States will need to maintain the ability to find and target these cells before they attack.

Second, the United States will face a *gray-zone fight*. Definitions vary regarding what the gray zone actually consists of, but China, Russia, Iran, and North Korea all have sought to achieve national objectives by using coercion short of armed conflict, often by exploiting U.S. and allied thresholds for response.[38] In many cases, this involves the use of covert, civilian, or proxy forces—such as China's People's Maritime Militia, Russia's "little green men," or Iran's Quds Force.[39]

---

[37] National Intelligence Council, 2017, p. 225.

[38] For a discussion, see Linda Robinson, Todd C. Helmus, Raphael S. Cohen, Alireza Nader, Andrew Radin, Madeline Magnuson, and Katya Migacheva, *Modern Political Warfare: Current Practices and Possible Responses*, Santa Monica, Calif.: RAND Corporation, RR-1772-A, 2017, pp. 2–6.

[39] For a sampling of the discussion, see Heidi Reisinger and Aleksandr Golts, *Russia's Hybrid Warfare: Waging War Below the Radar of Traditional Collective Defence*, Rome: NATO Defense College, Research Paper No. 105, November 2014; Paul Bucala and Frederick W.

Table 5.4
Four Types of Conflict

Category	Trend	Counterterrorism	Gray-Zone Fight	Asymmetric Conflict with a Second-Tier Power	High-End Conflict with a Near Peer
Geopolitical trends	Terrorism, weak states, and proxy wars in Islamic world	√			
Military trends	Increasing near-peer conventional modernization and professionalization				√
	Selectively capable second-tier powers			√	
	Adversary use of gray-zone tactics		√		
	Weakening of the state's monopoly on violence	√			
	AI as a class of disruptive technologies				√
Space/nuclear trends	Space as an increasingly contested environment				√
	Resumption of nuclear proliferation			√	
	Erosion of norms or treaties constraining tactical nuclear weapons use				√
Cyber trends	Increasing cyber espionage		√	√	√
	Increasing cyber sabotage		√	√	√
Restraint	Widespread distribution of imagery of military operations	√	√		
	Increasing public concern for civilian casualties	√	√		
	The spread of lawfare	√	√		
	Increasing power of false accusations	√	√		

Other times, gray-zone fights do not involve kinetic force at all, instead relying on information warfare, economic coercion, or cyber tools to achieve national objectives without provoking a shooting war. These strategies can be cheaper in blood and treasure than more-conventional forms of conflict, and they have been successful in many cases: Iran has used these tools to advance its interests in the Middle East, as have Russia in Eastern Europe and China in the South China Sea.

Looking forward to 2030, we can expect gray-zone conflicts to increase for two reasons. First, these tools have already proven themselves relatively cheap and successful, giving little reason for U.S. adversaries to stop using them. Second, the United States might be more vulnerable to this form of warfare, some of its allies perhaps even more so. As highlighted in the restraint and cyber trends in Table 5.4, the United States and its liberal democratic allies will be increasingly susceptible to lawfare, false accusations, and cyberattacks. They also will be harder pressed to control the narrative behind military operations, potentially creating more vulnerabilities that U.S. adversaries can exploit through gray-zone conflict. The combination of a proven track record of success and greater potential vulnerabilities could make U.S. adversaries more likely to turn to these tactics in the future.

Third, the United States could face an *asymmetric fight*, especially if the United States opts for overt military confrontation with either Iran or North Korea over the next decade. As we have mentioned, neither of those powers can match U.S. military power; instead, the United States will face numerically large but qualitatively inferior conventional forces. The true military challenges of these adversaries rest in their niche capabilities, in particular their anti-air and anti-ship missiles and their ability to use missiles, weapons of mass destruction, and cyber weapons in attacks on primarily nonmilitary targets.[40] Assum-

---

Kagan, *Iran's Evolving Way of War: How the IRGC Fights in Syria*, Critical Threats Project, March 15, 2016; and Ryan D. Martinson, "The Arming of China's Maritime Frontier," *China Maritime Report*, No. 2, Newport, R.I.: Naval War College, June 2017.

[40] For a good summary of the challenges involved in waging war with North Korea, see Michael J. Mazarr, Gian Gentile, Dan Madden, Stacie L. Pettyjohn, and Yvonne K. Crane, *The Korean Peninsula: Three Dangerous Scenarios*, Santa Monica, Calif.: RAND Corporation, PE-262-A, 2018.

ing both countries continue along their current modernization paths, the need for the United States to prepare for this asymmetric fight will remain unchanged in 2030 from today.

Finally, should the United States find itself in an overt conflict with either China or Russia, it will face a *high-end fight*. Of course, not all high-end fights are the same: A war against China in the maritime-centric Indo-Pacific will look different from a land-based conflict against Russia in Europe. That said, there are similarities. Although neither China nor Russia fully replicates the capabilities of the U.S. military, both will likely have homefield advantage, localized numerical superiority, and a nearing qualitative edge—complete with sophisticated weapon systems, such as advanced air defense systems, extensive offensive and defensive space and cyber capabilities, and the ability to wage a tactical nuclear war.[41] In sum, unlike any of the previous scenarios, these adversaries will be able in a high-end fight to contest all three domains—air, space, and cyber—where the USAF operates.

The trends outlined in Table 5.4 suggest that chances of high-end conflict, although still not likely, will probably increase by 2030. By modernizing and professionalizing their conventional forces, investing in cutting-edge technologies (such as cyber, offensive, space, and AI), and chipping away at the constraints on nuclear weapons (particularly in Russia's case), China and Russia will likely be better positioned a decade from now to wage high-end conflict with the United States. Of course, capability does not equate to motivation, and the logic of deterrence still applies, on balance. Nonetheless, the risks of a high-end conflict will likely increase over the next decade, if for no other reason than China and Russia will have more ability to wage such a conflict.

Ultimately, the emergence of these four archetypes of conflicts will pull the USAF and joint force at large in different directions. In some cases, the capabilities needed for each of these conflicts overlap,

---

[41] For a comparison, see Heginbotham et al., 2015; Scott Boston, Michael Johnson, Nathan Beauchamp-Mustafaga, and Yvonne K. Crane, *Assessing the Conventional Force Imbalance in Europe: Implications for Countering Russian Local Superiority*, Santa Monica, Calif.: RAND Corporation, RR-2402, 2018.

but in many others, they do not. For example, all four fights require ISR, but the high-end and asymmetric fights might privilege stealth and speed while the counterterrorism and gray-zone fights might require slower, more-persistent platforms. Similarly, the asymmetric fight might privilege the ability to quickly find and neutralize adversaries' stockpiles of missiles and weapons of mass destruction, but that same capability might be destabilizing to overarching strategic deterrence in a high-end fight in which Russia and China worry about their second-strike capabilities. There are similar trade-offs in terms of training and posture. Although the 2018 *National Defense Strategy* seemingly privileges training for the high-end fight against a major power, this is not because the other scenarios have gone away or even are less likely. Rather, this approach reflects conscious choices that current and future defense leaders will need to make.[42]

How Will the United States Fight? Declining Quantitative and Qualitative Advantage

Closely related to the question of what future conflict looks like is the question of how the USAF and the joint force at large will need to fight. As described in detail in the previous chapter, any number of trends will affect how the USAF fights and what capabilities it will need. Although the specific list of capabilities that the USAF and the joint force will need is important, the overall trend analysis points to a macro shift in the fundamental assumptions of how the United States wins wars—a shift away from the quantitative and qualitative advantages to which it has grown accustomed (Table 5.5).

During World War II, the United States relied on a quantitative advantage to propel it to victory. Its large defense industrial base allowed it to serve as the "arsenal of democracy,"[43] allowing it to pro-

---

[42] DoD, 2018, pp. 6, 9.

[43] Franklin D. Roosevelt, "Fireside Chat," radio address, December 29, 1940, transcript via American Presidency Project.

Table 5.5
Declining Quantitative and Qualitative Advantage

Category	Trend	Effect on U.S. Military Qualitative Superiority	Effect on U.S. Military Quantitative Superiority
Geopolitical trends	U.S. polarization and gridlock	Negative	Negative
Military trends	Decreasing U.S. conventional force size		Negative
	Increasing near-peer conventional modernization and professionalization	Negative	
	Selectively capable second-tier powers	Possibly negative	Possibly positive
	AI as a class of disruptive technologies	Negative	
Space/nuclear trends	Space as an increasingly contested environment	Negative	
	Proliferation of commercial space capabilities		Positive
	Resumption of nuclear proliferation	Negative	
	Erosion of norms and treaties constraining tactical nuclear weapons use	Negative	
Cyber trends	Increasing cyber espionage	Negative	
	Increasing cyber sabotage	Negative	
Global economic trends	Relatively declining U.S. and allied economic might	Negative	Negative
	The shrinking defense industrial base	Negative	Negative

duce equipment for not only itself but also its allies.[44] On a ship-for-ship, plane-for-plane, or tank-for-tank basis, the U.S. forces were equal, or in some cases inferior, to their Axis counterparts, but the United States could prevail based on its numerical advantages alone.[45]

During the Cold War, by contrast, the United States pinned its hopes on maintaining a qualitative advantage against the Soviet Bloc, at first relying on nuclear weapons to offset the Soviet conventional superiority in Europe.[46] Later, the United States developed—if unwittingly—a basket of technologies, such as precision weapons, night vision, and global positioning to offset the Soviet numerical advantages once again.[47] The United States demonstrated these technologies to great effect during the First Gulf War, but DoD admits those relative advantages are "eroding."[48] And although the term "third offset" went out of vogue with the change of presidential administrations, defense policymakers have been searching for new technological advantages.[49]

In 2030, the United States likely will not have a quantitative advantage, and its relative qualitative military edge might also decline, particularly in relation to China. As previously mentioned, the U.S. military is only a fraction of the size it was during the Cold War, and it is already smaller than the militaries of some its adversaries—most notably China—and likely will remain so until 2030. It is not at all clear that the United States could once again become that "arsenal of democracy" even if it were technologically feasible and politically palatable. As already

[44] Williamson Murray and Allen Read Millett, *A War to Be Won: Fighting the Second World War*, Cambridge, Mass.: Belknap Press, 2000, p. ix.

[45] For an overview of U.S. industrial production and its impact on the war effort, see Murray and Millett, 2000, pp. 533–540.

[46] Peter Grier, "The First Offset," *Air Force Magazine*, June 2016.

[47] Rebecca Grant, "The Second Offset," *Air Force Magazine*, July 2016.

[48] DoD, 2018, p. 1.

[49] For example, see Robert Work, Deputy Secretary of Defense, "The Third U.S. Offset Strategy and Its Implications for Partners and Allies," speech before the Center for a New American Security and the NATO Allied Command Transformation, Washington, D.C., January 28, 2015.

noted, the U.S. share of global GDP has been shrinking, although only slightly, while the share held by U.S. adversaries, particularly China, has been rising. Factoring in likely U.S. allies—such as the NATO countries, Japan, and other Asian allies—exacerbates this trend, because Europe and Japan have been losing global GDP far more rapidly than the United States. In 1990, the United States and its treaty allies made up almost 78.2 percent of the global economy; in 2017, that number was only 57.9 percent (Figure 5.1).

Future projections are highly dependent on model specifications, but a baseline projection—drawn from historical growth rates and current economic forecasts—suggests that China could very well become the largest single economy by 2030 and possibly by 2027, as measured by nominal GDP.[50] Factoring U.S. allies into the equation, the com-

[50] Options for projections included real GDP growth, growth of GDP in terms of purchasing-power parity, or the less-standard nominal GDP growth. None of these are the perfect indicator. We chose to use nominal GDP growth, which consists of real growth and inflation, because international purchases and investments take place in nominal dollars, so the share of nominal GDP gives one indication of international economic power. Real growth shows the growth in quantity of output, and growth of GDP in terms of purchasing-power parity is best used to show changes in the standard of living. The projection used the following simplifications. U.S. nominal GDP grew at an annual rate of 4.5 percent from 1990 to 2017; Japan's grew at a consistent 1.6 percent. The U.S. Congressional Budget Office projects annual nominal U.S. GDP growth of 4.1 percent from 2017 through 2028 (U.S. Congressional Budget Office, *The Budget and Economic Outlook: 2018 to 2028*, Washington, D.C., April 2018). Accordingly, we model U.S. future growth at 4.1 percent and Japanese future growth at 1.5 percent, the latter to take account of demographic changes. By contrast, the NATO 28 (excluding the United States) grew at 3.3 percent, but these are aging societies and growth has slowed, so we project forward at 2.0 percent. And U.S. allies in Asia have grown at 6.1 percent but have also recently slowed; they are also heavily dependent on China, which we expect to slow, so we project the Asian allies forward at 3.8 percent. Among challengers, China's annual nominal growth rate was 13.2 percent from 1990 to 2017 and 10.3 percent from 2010 to 2017, but this has been fueled in part by unsustainable debt and the Chinese leadership is intent on slowing growth, so we project forward at 9.0 percent. Russia's nominal GDP has been highly volatile and strongly related to oil and gas prices; it now faces a period of potentially low oil prices and continued sanctions. Accordingly, we project forward at 3.0 percent, below nominal growth from 2010 to 2017, a period that experienced both high oil prices and sanctions. The outlook for Iran is also related to oil and gas prices and sanctions, but Iran has more severe internal problems than Russia, poorer economic management, and will likely face tougher sanctions. Accordingly, we project forward at 2.0 percent. Finally, world nominal GDP growth has averaged 4.8 percent from 1990 to 2017, but 2.9 percent

Figure 5.1
Share of Nominal GDP of the United States, Allies, and Challengers, 1990–2017

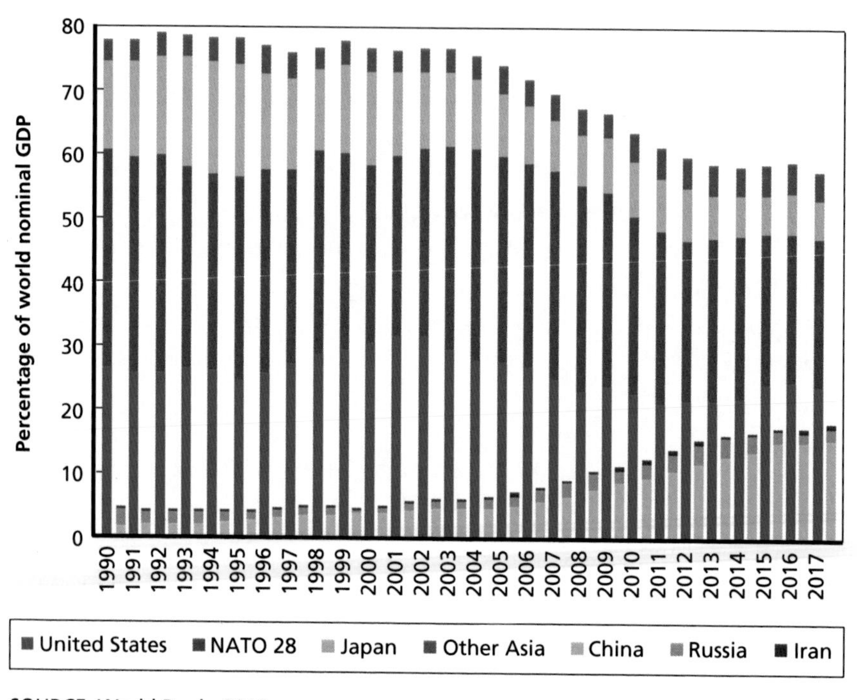

SOURCE: World Bank, 2018.
NOTES: GDP is in terms of nominal U.S. dollars. "NATO 28" consists of the 28 non-U.S. NATO members as of July 2018 (Albania, Belgium, Bulgaria, Canada, Croatia, Czech Republic, Denmark, Estonia, France, Germany, Greece, Hungary, Iceland, Italy, Latvia, Lithuania, Luxembourg, Montenegro, Netherlands, Norway, Poland, Portugal, Romania, Slovakia, Slovenia, Spain, Turkey, United Kingdom). "Other Asia" consists of U.S. treaty allies Australia, Korea, New Zealand, the Philippines, and Thailand. "China" includes Hong Kong and Macao because although the jurisdictions are under the one country–two systems model, it is likely that China would draw on Hong Kong and Macao as needed in wartime. The *National Security Strategy* (White House, 2017) identified China, Russia, Iran, and North Korea as challengers to the United States and its allies and to the global order. No GDP data were available for North Korea.

parison looks somewhat more favorable to the United States, although nowhere near the dominant position it enjoyed right after the Cold

---

since 2010. We project forward at 4.4 percent, meaning we project 5.0 percent growth for the rest of the world not included among the countries discussed.

War. By 2030, U.S. adversaries will constitute about 30 percent of the global economy, while the United States and its European and Asian allies will make up less than 50 percent (Figure 5.2). Most of this shift will come from the expected growth of China's share of global nominal GDP and the relative decline of the rest of NATO, particularly Europe. The United States will largely hold its own in the world economy, with 24.0 percent of world nominal GDP in 2017 and a projected 23.2 percent of world nominal GDP in 2030. In contrast, the share of world nominal GDP accounted for by the rest of NATO (the NATO 28) is projected to fall from 23.1 percent in 2017 to 17.1 percent in 2030. The share that Japan accounts for is also projected to fall, from 6.0 percent in 2017 to 4.2 percent in 2030.

Without a dominant economic base to support further military investments, it will become much more difficult for the United States to check the already existing trends of declining U.S. conventional military forces to counteract the military modernization and professionalization of near peers. The United States and its allies will need

Figure 5.2
Projections for Share of Nominal GDP Projections of the United States, Allies, and Challengers, 2017–2030

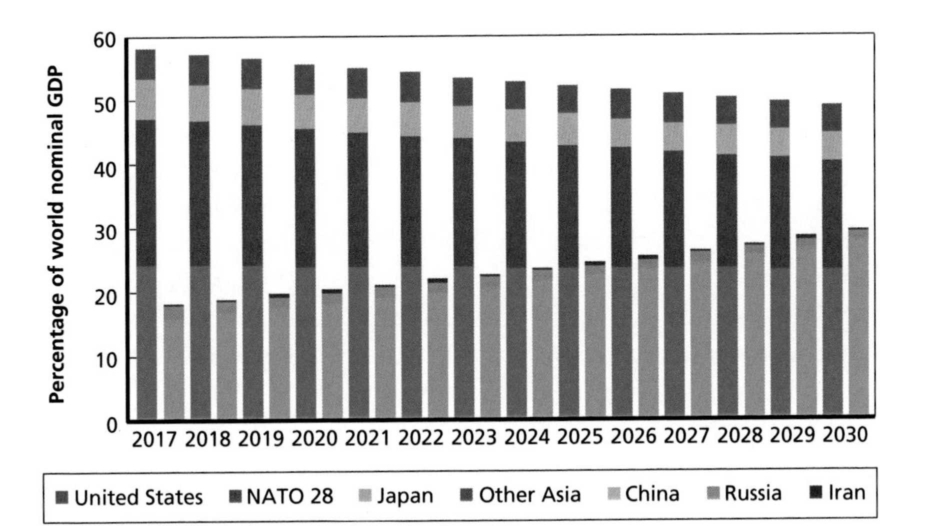

SOURCE: World Bank, 2018.

to devote larger portions of GDP to defense if they want to maintain a dominant military position. As noted in the geopolitical trends discussion in Chapter Four, the polarization and gridlock afflicting the United States are likely to hamper such sustained increases in defense. It is more probable that the United States' relative quantitative advantage will continue to decline. And these dynamics will likely be even more true among traditional U.S. allies in Europe, given their comparatively weaker economic circumstances.

The United States might not be able to rely on maintaining qualitative advantage going forward, either. As already noted, even second-tier adversaries are investing in ways—such as anti-satellite weapons, cyber espionage and sabotage weapons, and nuclear weapons—to blunt U.S. conventional military advantage.

If these asymmetric bets do not fully pay off, sustaining a qualitative military edge will depend in part on maintaining a healthy defense industrial base—with multiple companies competing to develop increasingly innovative weapon systems. However, the U.S. defense industrial base has shrunk. Of the top 100 global defense companies that existed in 1991, only 19 firms survived to 2014.[51] In 1960, there were 11 advanced military fixed-wing aircraft prime contractors in the United States; today, there are only three—Lockheed, Boeing, and Northrup Grumman (with General Atomics providing unmanned aerial systems).[52] In the near future, with production of fourth-generation fighter aircraft ending and no new fighter aircraft production slated to begin, the United States could find itself with only one type of fighter aircraft in active production—the F-35 joint strike fighter. Unsurprisingly, DoD's 2017 *Annual Industrial Capabilities Report to Congress* concluded, "The defense sector continues to financially outperform the broader U.S. equity market. . . . However, factors such as obsolescence, foreign dependency, fluctuating demand, industry consolidations, and loss of design teams and man-

---

[51] Censer, 2014.

[52] John Birkler, Paul Bracken, Gordon T. Lee, Mark A. Lorell, Soumen Saha, and Shane Tierney, *Keeping a Competitive U.S. Military Aircraft Industry Aloft: Findings from an Analysis of the Industrial Base*, Santa Monica, Calif.: RAND Corporation, MG-1133-OSD, 2011, p. 12.

ufacturing skills for critical defense products continue to threaten the health of the industrial base, limit innovation, and reduce U.S. competitiveness in the global markets."[53]

The ability of the United States to turn to its allies for help in compensating for these weaknesses will likewise remain uncertain at best. Although Europe retains a series of major defense companies—such as BAE Systems, Airbus Group (formerly EADS), and Thales—its defense industrial base remains fragmented and dominated by national defense companies, "leading to significant redundancies and inefficiencies in the regional industrial base."[54] European Union frameworks on defense affairs (such as the European Defence Agency, European Defence Fund, and Permanent Structured Cooperation) have not yet reduced these inefficiencies and generated promised innovation.[55] Similarly, U.S. allies in Asia have their own weaknesses. The Japanese defense industrial base, for example, struggles with cybersecurity and weak export controls that leave it vulnerable to foreign intelligence and commercial espionage.[56]

Finally, some of the key technologies for future combat—particularly in the cybersecurity and AI realms—might not come from the defense industrial base at all. Unlike previous military capabilities that were mostly by and for DoD, commercial and academic sectors are leading some of the most important research in these fields in the United States. China has invested significant state resources in these areas; so has Russia, although to a lesser extent.[57] Even if these tech-

---

[53] Office of the of the Under Secretary of Defense for Acquisition and Sustainment and Office of the Deputy Assistant Secretary of Defense for Manufacturing and Industrial Base Policy, *Fiscal Year 2017 Annual Industrial Capabilities Report to Congress*, Washington, D.C.: U.S. Department of Defense, March 2018, p. 2.

[54] Christina Balis and Henrik Heidenkamp, *Prospects for the European Defence Industrial Base*, London: Royal United Services Institute, Occasional Paper No. 9, September 2014, p. 4.

[55] These observations stem from interviews with government officials and researchers in London, Brussels, Berlin, and Warsaw in April 2018.

[56] Interview with defense reporter, Tokyo, June 18, 2018.

[57] Elsa Kania, "Beyond CFIUS: The Strategic Challenge of China's Rise in Artificial Intelligence," *Lawfare*, June 20, 2017; President of the Russian Federation, "Presentation of Era Innovation Technopolis," Moscow: Kremlin, February 23, 2018; Samuel Bendett, "Russia

nologies are developed by U.S. or allied companies and universities, it is still possible that adversaries could buy or steal the information.[58] As a result, there is no guarantee that the U.S. government will have a monopoly on the use of any innovations or that the joint force will be able to maintain a qualitative edge over its adversaries.

When and Why Will the United States Fight? Increasing Inability to Control Strategic Outcomes

Finally, there are two intertwined questions of when and why the United States will fight wars. Perhaps no single question in the study of international relations has attracted as much attention as why wars occur, and yet, for all that effort—and as Secretary Gates' remark captures—the United States still seems surprised when wars break out. We cannot, in this report, predict with any fidelity when and why the next war will occur, but the early trends do indicate four overarching themes that inform and shape the answers to these questions (Table 5.6).

First, some of the traditional guardrails that the United States has relied on for the past several decades to prevent major power conflict seem to be eroding. As already mentioned, the overwhelming U.S. conventional military superiority is waning—resulting partly from decisions by the United States, actions by its adversaries, and shifts in the conduct of conflict itself—and with that shifts in conventional deterrents for future conflict. Similarly, on the economic front, the free-trade regime that promotes economic cooperation and minimizes the chances of states resorting to violence over access to markets is also under pressure. As U.S. and allied economic might decline in relative terms, and as the power of U.S.-led sanctions wanes concurrently, so could U.S. ability to coerce results without resorting to violence. Above all, if polarization and gridlock continue, the United States might find

---

Wants to Build a Whole City for Developing Deadly Weapons," *The National Interest*, March 29, 2018.

[58] For example, see Zach Dorfman, "How Silicon Valley Became a Den of Spies," *Politico*, July 27, 2018.

Table 5.6
An Increasing Inability to Control Strategic Outcomes

Category	Trend	Why the United States Will Fight	Declining Guardrails	Mounting Internal Pressures	Potential Entrapment	Potential Exogenous Shock
Geopolitical trends	U.S. polarization and retrenchment	Vacuum from U.S. leadership	√			
	Rising China	Rising ambitions		√		
	Growing tensions in Asia	Nationalism; fear of rising China			√	
	A revanchist Russia	Combination of Russian insecurity about West and ambitions		√		
	European turmoil	Insecurity about Russian aggression		√	√	
	Terrorism, weak states, and proxy wars in the Islamic world	Instability leading potential for exogenous shocks		√	√	√
Military trends	Decreasing U.S. conventional force size	Regional aggressor calculates that the United States lacks the capacity and resolve to respond effectively	√			
	Increasing near peer conventional modernization and professionalization	China or Russia calculates it can deny the United States sufficient access to defeat effort to change territorial status quo	√			
	Adversary use of gray-zone tactics	States victimized by covert or proxy forces will need support			√	
	Weakening of the state's monopoly on violence	States unable to restrain heavily armed individuals and groups will need support		√		√
	AI as a class of disruptive technologies	Regional aggressor believes its AI capabilities are sufficient to change the status quo	√			√

Table 5.6—Continued

Category	Trend	Why the United States Will Fight	Declining Guardrails	Mounting Internal Pressures	Potential Entrapment	Potential Exogenous Shock
Nuclear trends	Resumption of nuclear proliferation	Escalation of conventional crises or conflicts				√
Global economic trends	Increasing pressure on the global trading system	Fear of economic damage from closing markets	√	√		
	The rise of China	Miscalculation; Chinese ambitions or opportunism or a China in crisis will go to war to spur cohesion and nationalism	√	√	√	
	The search for new resources	To protect a partner under attack or to forestall monopolization of a resource			√	
	Declining U.S. and allied economic might	Miscalculation or opportunistic action to take advantage of perceived weakness	√			
Environmental trends	Rising temperatures	Reduced economic productivity could weaken governments and exacerbate ethnic tensions		√		√
	Sea level rise	Displacement from sea level rise will add to internal pressure on states		√		√
	Extreme weather events	Could increase spread of disease and migration instigating spikes in violence		√		√
	Water scarcity	More water-related intrastate and interstate conflict and unrest		√	√	√
	Urbanization and megacities	Increased chances for failed governance and lawlessness		√		√

it progressively harder to provide global leadership in maintaining the international order.

Second, internal pressures that can cause conflict might be mounting. Aside from the aforementioned problems of economic discontent, adverse environmental changes, further upending economic and societal structures, migration, and terrorism, the past several years in particular have seen the rise across the globe of the strongmen who will shape the next decade of politics. In some cases, such as Russian President Vladimir Putin's recent election to a fourth term or Chinese President Xi Jinping's decision to waive term limits, the strongmen are consolidating power at the helm of U.S. competitors. In other cases, such as Crown Prince Muhammad Bin Salman, these strongmen are at the helm of U.S. partners and allies. On the one hand, the likely longevity of these strongmen could act as a stabilizing force in the international system; on the other, it creates possible incentives for these leaders to secure major foreign policy accomplishments to cement their legacies and legitimize their rule at home.[59] Of more concern, though, is what would happen if these leaders ever felt their power wane. Rule by strongmen also functionally limits the safety valves for domestic discontent, focusing any grievances on the man at the top and creating a potentially explosive mix should that leader fall from power.[60] And any leader who could hold on might be willing to risk war to secure the domestic power base.[61]

Third, the potential for the United States to become entrapped in a regional conflict not of its choosing is also increasing. As mentioned, U.S. adversaries see much of the U.S.-led international order as antithetical to their interests and want to change it. Even if this does not

---

[59] For example, see Chris Buckley and Adam Wu, "Ending Term Limits for China's Xi Is a Big Deal. Here's Why," *New York Times*, March 10, 2018; Andrew E. Kramer, "For Putin's 4th Term, More a Coronation Than an Inauguration," *New York Times*, May 7, 2018.

[60] For example, this concern was repeatedly expressed about Muhammad Bin Salman in Saudi Arabia. Drawn from interviews with government officials, think tank analysts, and academics, Tel Aviv, Amman, and Abu Dhabi, May 8–15, 2018.

[61] Most notably, this concern is expressed about Xi's willingness to resort to war as China's economy slows. Drawn from interviews with government officials, think tank analysts, and academics, Beijing, June 12–15, 2018.

directly lead to conflict with the United States, U.S. allies have sensitivities that differ from the United States. Poland and the Baltic states are understandably more sensitive to Russian actions on their borders. Israel views Iranian military presence in Syria as an existential threat and has responded with force to prevent Iran from cementing its military foothold in the region.[62] Japan considers the Senkaku Islands—also claimed by China—as part of its sovereign territory and Taiwan as integral to its defense of its southern islands.[63] Especially because gray-zone warfare involves pressing on these sensitive areas while avoiding overt uses of force, the United States might find itself confronting an entrapment problem in 2030. If China, Russia, Iran, or North Korea trample over a U.S. ally's redlines, the United States could be faced with the difficult choice of entering into a war it does not want or abandoning an ally.

Finally, the number of exogenous shocks that could spark conflict in the years to come also might increase. There are many traditional catalysts for conflict. Terrorism, particularly in the Middle East, remains a concern and (as shown by the attacks in 2001) can spark large-scale intervention. The resumption of nuclear proliferation, particularly by Iran and North Korea, also could serve as a catalyst for war. There are also increasing opportunities for missteps to lead to crises and perhaps outright conflict. U.S. military forces already bump up against Chinese forces in such places as the South China Sea and against Russian forces in Syria, creating chances for accidents and inadvertent escalation. As China's economic and security interests expand through the Belt and Road Initiative, and as Russia becomes more active in its near abroad, the number of potential flashpoints will increase.

There are also deeper, more systemic trends that could spark conflict. As the climate changes and the number of extreme weather events increases, so too do the chances of one of these events triggering inter-

---

[62] Interviews with government officials, think tank analysts, and academics, Tel Aviv, May 8–10, 2018.

[63] As one interviewee said, a Chinese takeover of Taiwan would be "game over for Japan." Interview with a Japanese academic, Tokyo, June 18, 2018. Other Japanese officials agreed with the sentiment, although in more measured terms. Interviews with government officials, think tank analysts, and academics, in Tokyo, June 18–19, 2020.

nal instability or cross-border migration flows.[64] Scant resources pose a similar danger. Currently, the search for scarce resources typically does not cause conflict among developed countries because it is usually cheaper to buy these resources on global markets than resort to conquest. This could change if the global trading system erodes.[65] And although the United States might become a net exporter of natural gas in the coming decade, it still depends on imports of oil and other natural resources to drive its economy.[66] Of the 33 minerals (or groups of minerals, such as rare-earth elements) that the Department of the Interior listed in February 2018 as essential to the U.S. economy, China was the top producer for 19 and the top supplier for 12.[67]

The result of these four trends is that the chances of a large-scale conflict by 2030 could increase, and the ability of the United States to control when and why it occurs is likely to decline. To be sure, the United States never had an absolute ability to dictate world events, even at the height of its relative power at the end of the Cold War—as the first Gulf War and September 11 terrorist attacks demonstrate. Nonetheless, the United States might find itself increasingly forced into conflict by aggression from an adversary or entrapment of an ally rather than dictating conflict at a time and place of its own choosing.

---

[64] World Bank, 2016.

[65] David G. Victor, "What Resource Wars?" *The National Interest*, November 12, 2007.

[66] U.S. Energy Information Administration, *Annual Energy Outlook 2017*, Washington, D.C.: U.S. Department of Energy, January 5, 2017.

[67] U.S. Department of the Interior, Office of the Secretary, "Draft List of Critical Minerals," *Federal Register*, Vol. 83, No. 33, February 16, 2018, pp. 7065–7068.

CHAPTER SIX

Implications for the U.S. Air Force and the Joint Force

As the future of warfare places more demands on the U.S. military force and pulls its limited resources in opposite directions, the United States will increasingly face a grand strategic choice. On the one hand, it can break with the internationalist foreign policy that it has pursued at least since the end of the Cold War and become dramatically more selective about where, when, and why it commits forces. On the other hand, it can maintain its current commitments, knowing full well that doing so will come at a significantly greater cost—at least in treasure and possibly in blood. Whatever choice the United States makes, this decision will not be made by the USAF or by DoD at large. Assuming the United States opts for the latter approach and decides to maintain its current commitments, this report offers some insights regarding how to shape the force in a general sense in terms of capability, capacity, posture, strategy, and overall policy.

Capability: Range, Precision, Information, and Automation

Although we focused more on the future drivers of conflict than on new technologies that might be developed over the next decade and a half, the USAF and the joint force should invest in additional capabilities in four general areas—range, precision, information, and automation.

First, future conflicts will likely place a premium on being able to operate at range. As China becomes more militarily formidable and more geopolitically assertive, the USAF will need to operate over the vast expanses of the Indo-Pacific. As even second-tier adversaries—such as North Korea and Iran—invest in anti-access area denial capabilities, the USAF will likely want to stay outside these adversaries' missile ranges. And as excessive heat, rising sea levels, or extreme weather make it more difficult to operate in certain areas of world, the USAF might find it necessary to base from afar.

Second, the trends in restraint and geography, the increasing salience of lawfare, the wider distribution of imagery of military operations, and the growing urbanization of the global population all indicate that the USAF and the joint force should invest in increasing their precision to avoid the legal and political backlash that comes with civilian casualties. It is an open question whether the United States would be as concerned about these issues in high-intensity conflict over its core security interests, but these restraints will almost certainly persist in counterterrorism and counter–gray zone operations—and the latter shows no sign of going away.

Third, particularly given the trends in cyber and gray-zone conflict, the USAF and joint force will need to enhance their information warfare capabilities. Particularly as gray-zone operations become central to great-power competition and as cyber becomes an increasingly valuable tool of espionage, sabotage, and subversion, information will increasingly be used as a weapon and considered a domain of warfare. Waging wars successfully in this domain will require not only a renewed emphasis on psychological operations and cyber but also a rethinking of the role of other parts of the services, such as the public affairs and judge advocate general communities.

Finally, because of the trend toward greater use of AI, the USAF and the joint force will need to invest in automation. Although no one can be certain just how AI will shape the conduct of warfare, it is certain that it will have a profound effect—from speeding up the pace of the targeting cycles to changing the very nature of when and how wars are fought in the first place.

Capacity: "More"

This project was not designed to identify and did not yield precise numbers on how large either the joint force or the USAF will need to be to fight future wars. However, the preceding analysis suggests that the force will probably need to be larger than it is today.[1] The joint force and USAF are currently a fraction of their size during the Cold War, the last major period of strategic competition. The platforms today are arguably more capable than they were a quarter-century ago, but as former Secretary of the Navy Ray Mabus argued, "quantity has a quality all its own if you want capability."[2]

Mabus was channeling an adage that will likely find new meaning in 2030. As discussed in this report, the United States will face at least five credible adversaries—including two near peers—in four different types of conflict spread through at least three different geographical regions of the world. Platforms can be in only one place at one time and service members can specialize in only a finite number of missions, with the net result being that the joint force at large—and the USAF in particular (because air, space, and cyber will likely play crucial roles in any conflict)—will be increasingly stretched. Although DoD is experimenting with how to more efficiently train and employ the resources it has through such concepts as "dynamic force employment," these measures go only so far.[3] The capacity problem is arguably more acute when it comes to munitions. In 2016, the USAF raided its stockpiles of precision munitions to resource the air campaign against the Islamic State, a campaign that presumably called for only a fraction of the demand required for a high-intensity conflict against a single

[1] For an attempt to quantify the question, see Jacob Cohn, Ryan Boone, and Thomas G. Mahnken, *How Much Is Enough? Alternative Defense Strategies*, Washington, D.C.: Center for Strategic and Budgetary Assessments, 2016; David Ochmanek, Peter A. Wilson, Brenna Allen, John Speed Meyers, and Carter C. Price, *U.S. Military Capabilities and Forces for a Dangerous World: Rethinking the U.S. Approach to Force Planning*, Santa Monica, Calif.: RAND Corporation, RR-1782-RC, 2017.

[2] Ray Mabus, remarks at 27th annual Emerging Issues Forum: Investing in Generation Z, Raleigh, N.C.: February 7, 2012.

[3] DoD, 2018, p. 5

more-sophisticated state adversary, let alone the possibility of multiple simultaneous conflicts.[4] Indeed, outside analyses regularly conclude that "[p]rogrammed stocks of standoff and other preferred munitions are seriously inadequate for wartime needs."[5]

If, as the *National Defense Strategy* claims and our study reaffirms, we have entered a period of strategic competition with five credible adversaries (including two near peers), with at least four distinct types of conflict and three theaters of concern, the United States will need to size and arm its force accordingly.

Posture: Few Gains to Be Realized by Drawing Down in the Middle East

On a related note, the future of warfare will likely continue to demand a robust forward posture in all three regions of concern, including the Middle East. Despite the stated desire of the *National Defense Strategy* to focus on interstate competition rather than on terrorism, and despite the stated desire of multiple administrations to extricate the United States from the Middle East, the region remains the most likely—though not the most dangerous—place where the United States will need to fight wars in the future.[6] This assessment has important implications for the USAF because of its increasingly central role in counterterrorism missions, thanks in part to the trends in U.S. restraints on the use of force and the continued public aversion to using ground forces in this region. The USAF will be unlikely to shift many assets out of the Middle East to support seemingly higher-priority missions in the Indo-Pacific and in Europe. At the same time, as extreme heat makes it harder to fly from air bases in the Gulf states, the USAF will need to rethink how it operates in the region. Finally, it is likely that

---

[4] Marcus Weisgerber, "The US Is Raiding Its Global Bomb Stockpiles to Fight ISIS," *Defense One*, May 26, 2016.

[5] David Ochmanek, *Restoring U.S. Power Projection Capabilities: Responding to the 2018 National Defense Strategy*, Santa Monica, Calif.: RAND Corporation, PE-260-AF, 2018, p. 9.

[6] DoD, 2018, p. 1.

the USAF will continue to need forces for these "low-end" fights (e.g., light attack aircraft and non-stealthy ISR) while it still prepares for high-end fights against Russia and China.

Strategy: The Growing Need for Agility at All Levels

The need for flexibility, adaptability, and agility have become so common in discussions of the future of warfare as to make their recommendation seem almost hackneyed. Indeed, all three terms run throughout both the *National Defense Strategy* and the USAF's 2015 *Strategic Master Plan*, with the latter labeling "agility" as one of two imperatives for the service.[7] These concepts remain a catchphrase, yet the need for agility at all levels is real. On the policy level, the U.S. alliance structure will need to adapt as Europe becomes increasingly fragmented and preoccupied with its own issues. The United States will also need to be nimble enough to capitalize on opportunities to strengthen and build new alliances and partnerships in Asia that will be created by the growing threat of a rising China. On a strategic level, the joint force will need to be able to adapt to changing environments more quickly and to shift assets more readily across theaters and problem sets in response to adversaries' actions. Operationally and tactically, the joint force will need to be able to shift focus across the range of conflict and adapt to new technologies, including those that can credibly threaten areas where the United States has enjoyed virtual supremacy, such as the air and space domains. In sum, as U.S. quantitative and qualitative military advantages diminish, and as it becomes more difficult for the United States to control strategic outcomes, strategic agility will become a necessary cornerstone for the force of 2030.

---

[7] DoD, 2018; U.S. Air Force, *USAF Strategic Master Plan*, Washington, D.C., May 2015.

Policy: Increasing Resilience on the Home Front

Winning the next war depends only partially on developing the right capabilities with sufficient capacity postured in the right places with the right strategies to employ them. It also requires maintaining the economic wherewithal and the political will to sustain and prevail in such a conflict, especially for those fought against rival great powers. Successfully preparing for the future of warfare remains only partially within the control of the USAF or even DoD. Indeed, as previously argued, one reason why future conflict could occur is because the liberal economic order that the United States traditionally upheld might be eroding and one reason the United States finds itself in the geostrategic position it is in today is because of internal polarization and gridlock. The key to winning the next war, then, might lie only partially within the annals of the defense budget and to a greater extent with U.S. policymakers—and, perhaps, the U.S. public at large.

The Future of Warfare: A Deepening Series of Strategic Dilemmas

What is the future of warfare in 2030? There is no single answer to the question. It cannot be reduced to the growth of artificial intelligence, the rise of China, or the decline of the liberal economic order—no matter how significant each of these individual trends might be. Indeed, the central lesson of how to avoid falling into the trap of so many previous failed attempts to predict the future of warfare might be to not focus single-mindedly on any one trend in any specific area.

The future of warfare similarly defies any singular historical analogy.[8] There will be multiple great powers vying for influence, as occurred in the run-up to World War I; an erosion of free trade and international institutions, as preceded World War II; the presence of nuclear weapons to limit the conflict, as happened in the Cold War;

[8] For another exploration of this topic, see Mazarr et al., 2018.

ongoing Islamic terrorism, as defined the early 2000s, and a host of other trends that the world has never seen before.

If there is a single way to describe the future of warfare, it is as a deepening series of strategic dilemmas: between preparing for the low end of the spectrum of conflict and the high one, between planning for the wars that the United States most likely will fight and the ones it most hopes to avoid, between maintaining current U.S. allies and cultivating new ones. On top of all this is the necessity of making a finite amount of resources go farther in a future with ever fewer strategic certainties.

References

Allison, Graham, *Destined for War: Can America and China Escape Thucydides's Trap?* New York: Houghton Mifflin Harcourt, 2017.

———, "What Xi Jinping Wants," *The Atlantic*, May 31, 2017. As of July 24, 2018:
https://www.theatlantic.com/international/archive/2017/05/what-china-wants/528561/

Auslin, Michael, "Duterte's Defiance," *Foreign Affairs*, November 2, 2016.

Balis, Christina, and Henrik Heidenkamp, *Prospects for the European Defence Industrial Base*, London: Royal United Services Institute, Occasional Paper No. 9, September 2014.

Bendett, Samuel, "Russia Wants to Build a Whole City for Developing Deadly Weapons," *The National Interest*, March 29, 2018. As of May 30, 2018:
http://nationalinterest.org/blog/the-buzz/russia-wants-build-whole-city-developing-deadly-weapons-25121

Biddle, Stephen, "Afghanistan and the Future of Warfare," *Foreign Affairs*, Vol. 82, No. 2, March–April 2003, pp. 31–46.

Biddle, Stephen, and Jeffrey A. Friedman, *The 2006 Lebanon Campaign and the Future of Warfare: Implications for Army and Defense Policy*, Carlisle, Pa.: Strategic Studies Institute, 2008.

Birkler, John, Paul Bracken, Gordon T. Lee, Mark A. Lorell, Soumen Saha, and Shane Tierney, *Keeping a Competitive U.S. Military Aircraft Industry Aloft: Findings from an Analysis of the Industrial Base*, Santa Monica, Calif.: RAND Corporation, MG-1133-OSD, 2011. As of February 14, 2019:
https://www.rand.org/pubs/monographs/MG1133.html

Boston, Scott, Michael Johnson, Nathan Beauchamp-Mustafaga, and Yvonne K. Crane, *Assessing the Conventional Force Imbalance in Europe: Implications for Countering Russian Local Superiority*, Santa Monica, Calif.: RAND Corporation, RR-2402, 2018. As of July 27, 2018:
https://www.rand.org/pubs/research_reports/RR2402.html

Boston, Scott, and Dara Massicot, *The Russian Way of Warfare: A Primer*, Santa Monica, Calif.: RAND Corporation, PE-231-A, 2017. As of July 5, 2018: https://www.rand.org/pubs/perspectives/PE231.html

Bryce Space and Technology, "State of the Satellite Industry Report," Alexandria, Va., for the Satellite Industry Association, June 2017. As of August 7, 2018: https://www.sia.org/wp-content/uploads/2017/07/SIA-SSIR-2017.pdf

Bucala, Paul, and Frederick W. Kagan, *Iran's Evolving Way of War: How the IRGC Fights in Syria*, Critical Threats Project, March 15, 2016. As of May 21, 2018: https://www.criticalthreats.org/analysis/irans-evolving-way-of-war-how-the-irgc-fights-in-syria

Buckley, Chris, and Adam Wu, "Ending Term Limits for China's Xi Is a Big Deal. Here's Why," *New York Times*, March 10, 2018. As of July 27, 2018: https://www.nytimes.com/2018/03/10/world/asia/china-xi-jinping-term-limit-explainer.html

Canton, James, "The Extreme Future of Megacities," *Significance*, Vol. 8, No. 2, 2011, pp. 53–56.

Censer, Marjorie, "Defense Companies Brace for a Different Kind of Consolidation This Time Around," *Washington Post*, January 12, 2014. As of July 30, 2018: https://www.washingtonpost.com/business/capitalbusiness/defense-companies-brace-for-a-different-kind-of-consolidation-this-time-around/2014/01/10/a38e5152-5dec-11e3-bc56-c6ca94801fac_story.html?utm_term=.a07fdf211f45

Cheng, Jonathan, "North Korea Expands Key Missile-Manufacturing Plant," *Wall Street Journal*, July 1, 2018. As of August 2, 2018: https://www.wsj.com/articles/north-korea-expands-key-missile-manufacturing-plant-1530486907

Cheng Xiaobo, ed. [程晓波主编], *Big Data Report on Trade Cooperation Under the Belt and Road Initiative* [一带一路贸易合作大数据报告], "Belt and Road" Big Data Center of the State Information Center [国家信息中心"一带一路"大数据中心] and SINOIMEX [大连瀚闻资讯有限公司], May 2018.

Chivvis, Christopher S., Raphael S. Cohen, Bryan Frederick, Daniel S. Hamilton, F. Stephen Larrabee, and Bonny Lin, *NATO's Northeastern Flank: Emerging Opportunities for Engagement*, Santa Monica, Calif.: RAND Corporation, RR-1467-AF, 2017. As of July 2, 2018: https://www.rand.org/pubs/research_reports/RR1467.html

Coats, Daniel R., *Statement for the Record: Worldwide Threat Assessment of the U.S. Intelligence Community*, Washington, D.C.: Office of the Director of National Intelligence, February 13, 2018. As of August 6, 2018: https://www.dni.gov/files/documents/Newsroom/Testimonies/2018-ATA---Unclassified-SSCI.pdf

Cohen, Eliot A., and John Gooch, *Military Misfortunes: The Anatomy of Failure in War*, New York: Free Press, 1990.

Cohen, Raphael S., *The History and Politics of Defense Reviews*, Santa Monica, Calif.: RAND Corporation, RR 2278-AF, 2018. As of June 6, 2018: https://www.rand.org/pubs/research_reports/RR2278.html

Cohen, Raphael S., Eugeniu Han, and Ashley L. Rhoades, *Geopolitical Trends and the Future of Warfare: The Changing Global Environment and Its Implications for the U.S. Air Force*, Santa Monica, Calif.: RAND Corporation, RR-2849/2, 2020. As of May 2020: https://www.rand.org/pubs/research_reports/RR2849z2.html

Cohen, Raphael S., David E. Johnson, David E. Thaler, Brenna Allen, Elizabeth M. Bartles, James Cahill, and Shira Efron, *From Cast Lead to Protective Edge: Lessons from Israel's Wars in Gaza*, Santa Monica, Calif.: RAND, RR-1888-A, 2017. As of July 30, 2018: https://www.rand.org/pubs/research_reports/RR1888.html

Cohn, Jacob, Ryan Boone, and Thomas G. Mahnken, *How Much Is Enough? Alternative Defense Strategies*, Washington, D.C.: Center for Strategic and Budgetary Assessments, 2016. As of July 26, 2018: https://csbaonline.org/research/publications/how-much-is-enough-alternative-defense-strategies

Cronin, Patrick M., Richard Fontaine, Zachary M. Hosford, Oriana Skylar Mastro, Ely Ratner, and Alexander Sullivan, *The Emerging Asia Power Web: The Rise of Bilateral Intra-Asian Security Ties*, Washington, D.C.: Center for a New American Security, 2013. As of November 16, 2017: https://www.files.ethz.ch/isn/165406/CNAS_AsiaPowerWeb.pdf

Department of the Air Force, *Air Force Strategic Environment Assessment 2014–2034*, Washington, D.C., 2015.

DoD—*See* U.S. Department of Defense.

Dorfman, Zach, "How Silicon Valley Became a Den of Spies," *Politico*, July 27, 2018. As of July 27, 2018: https://www.politico.com/magazine/story/2018/07/27/silicon-valley-spies-china-russia-219071

Dunlap, Jr., Charles J., "Lawfare: A Decisive Element of 21st-Century Conflicts?" *Joint Forces Quarterly*, No. 54, Third Quarter 2009.

Economy, Elizabeth C., "The Great Firewall of China: Xi Jinping's Internet Shutdown," *The Guardian*, June 29, 2018. As of August 7, 2018: https://www.theguardian.com/news/2018/jun/29/the-great-firewall-of-china-xi-jinpings-internet-shutdown

Efron, Shira, Kurt Klein, and Raphael S. Cohen, *Environment, Geography and the Future of Warfare: The Changing Global Environment and Its Implications for the U.S. Air Force*, Santa Monica, Calif.: RAND Corporation, RR-2849/5-AF, 2020. As of May 2020:
https://www.rand.org/pubs/research_reports/RR2849z5.html

Evenett, Simon J., and Johannes Fritz, *Going Spare: Steel, Excess Capacity, and Protectionism*, 22nd Global Trade Alert Report, London: CEPR Press, 2018.

Ferdinando, Lisa, "Carter Outlines Security Challenges, Warns Against Sequestration," DoD News, Defense Media Activity, March 17, 2016. As of October 17, 2017:
https://www.defense.gov/News/Article/Article/696449/carter-outlines-security-challenges-warns-against-sequestration/

Frederick, Bryan, and Nathan Chandler, *Restraint and the Future of Warfare: The Changing Global Environment and Its Implications for the U.S. Air Force*, Santa Monica, Calif.: RAND Corporation, RR-2849/6, 2020. As of May 2020:
https://www.rand.org/pubs/research_reports/RR2849z6.html

Gallup, "Confidence in Institutions," 2017. As of November 8, 2017:
http://news.gallup.com/poll/1597/Confidence-Institutions.aspx

Gates, Robert, speech to the U.S. Military Academy, West Point, N.Y., February 25, 2011. As of June 5, 2017:
http://archive.defense.gov/Speeches/Speech.aspx?SpeechID=1539

Gleick, Peter H., "Water, Drought, Climate Change, and Conflict in Syria," *Weather, Climate, and Society*, Vol. 6, No. 3, July 2014, pp. 331–340.

Graham, David A., "The Fragile Future of Reform in Saudi Arabia," *The Atlantic*, June 27, 2018. As of June 29, 2018:
https://www.theatlantic.com/international/archive/2018/06/the-fragile-future-of-reform-in-saudi-arabia/563846/

Grant, Rebecca, "The Second Offset," *Air Force Magazine*, July 2016, pp. 32–36. As of July 25, 2018:
http://www.airforcemag.com/MagazineArchive/Pages/2016/July%202016/The-Second--Offset.aspx

Grier, Peter "The First Offset," *Air Force Magazine*, June 2016, pp. 56–60. As of July 30, 2018:
http://www.airforcemag.com/MagazineArchive/Magazine%20Documents/2016/June%202016/0616offset.pdf

Heginbotham, Eric, Michael Nixon, Forrest E. Morgan, Jacob L. Heim, Jeff Hagen, Sheng Li, Jeffrey Engstrom, Martin C. Libicki, Paul DeLuca, David A. Shlapak, David R. Frelinger, Burgess Laird, Kyle Brady, and Lyle J. Morris, *The U.S.-China Military Scorecard: Forces, Geography, and the Evolving Balance of Power, 1996–2017*, Santa Monica, Calif.: RAND Corporation, RR-392-AF, 2015. As of July 5, 2018:
https://www.rand.org/pubs/research_reports/RR392.html

Herring, David, "Climate Change: Global Temperature Projections," webpage, March 6, 2012. As of March 4, 2018:
https://www.climate.gov/news-features/understanding-climate/climate-change-global-temperature-projections

Heydarian, Richard Javad, "Duterte's Dance with China," *Foreign Affairs*, April 26, 2016. As of July 2, 2018:
https://www.foreignaffairs.com/articles/philippines/2017-04-26/dutertes-dance-china

Howard, Michael, "Men Against Fire: Expectations of War in 1914," *International Security*, Vol. 9, No. 1, Summer, 1984, pp. 41–57.

International Institute for Strategic Studies, *The Military Balance*, Vol. 118, No. 1, 2018.

Jackson, Matthew O., and Massimo Morelli, "The Reasons for Wars: An Updated Survey," in Christopher J. Coyne and Rachel L. Mathers, eds., *The Handbook on the Political Economy of War*, Cheltenham, UK, and Northampton, Mass.: Edward Elgar, 2011, pp. 34–57.

Jefferies & Co., Global Technology, Media, and Telecom Team, "Mobility 2020: How an Increasingly Mobile World Will Transform TMT Business Models over the Coming Decade," New York, September 2011. As of May 3, 2018:
http://www.slideshare.net/allabout4g/mobility-2020

Joint Chiefs of Staff, *Joint Operating Environment (JOE) 2035: The Joint Force in a Contested World*, Washington, D.C., July 14, 2016.

Kania, Elsa, "Beyond CFIUS: The Strategic Challenge of China's Rise in Artificial Intelligence," *Lawfare*, June 20, 2017. As of May 30, 2018:
https://www.lawfareblog.com/beyond-cfius-strategic-challenge-chinas-rise-artificial-intelligence

Katzenstein, Peter J., ed., *The Culture of National Security: Norms and Identity in World Politics*, New York: Columbia University Press, 1996.

Katzman, Kenneth, *Iran: Politics, Human Rights, and U.S. Policy*, Washington, D.C.: Congressional Research Service, RL32048, May 21, 2018. As of July 24, 2018:
https://fas.org/sgp/crs/mideast/RL32048.pdf

Kavanagh, Jennifer, and Michael D. Rich, *Truth Decay: An Initial Exploration of the Diminishing Role of Facts and Analysis in American Public Life*, Santa Monica, Calif.: RAND Corporation, RR-2314-RC, 2018. As of July 30, 2018: https://www.rand.org/pubs/research_reports/RR2314.html

Kramer, Andrew E., "For Putin's 4th Term, More a Coronation Than an Inauguration," *New York Times*, May 7, 2018. As of July 27, 2018: https://www.nytimes.com/2018/05/07/world/europe/putin-inauguration-russia-president.html

Kraska, James, *How China Exploits a Loophole in International Law in Pursuit of Hegemony in East Asia*, Philadelphia, Pa.: Foreign Policy Research Institute, January 22, 2015.

Krepinevich, Jr., Andrew F., *The Army and Vietnam*, Baltimore, Md.: Johns Hopkins University Press, 1986.

Kronstadt, K. Alan, and Shayerah Ilias Akhtar, *India-U.S. Relations: Issues for Congress*, Washington, D.C.: Congressional Research Service, R44876, June 19, 2017. As of November 22, 2017: https://fas.org/sgp/crs/row/R44876.pdf

Larson, Eric V., and Bogdan Savych, *Misfortunes of War: Press and Public Reactions to Civilian Deaths in Wartime*, Santa Monica, Calif.: RAND Corporation, MG-441-AF, 2007. As of July 30, 2018: https://www.rand.org/pubs/monographs/MG441.html

Mabus, Ray, remarks at 27th annual Emerging Issues Forum: Investing in Generation Z, Raleigh, N.C.: February 7, 2012. As of July 26, 2018: http://www.navy.mil/navydata/people/secnav/Mabus/Speech/emergingissuesfinal.pdf

Mansfield, Edward D., *Power, Trade, and War*, Princeton, N.J.: Princeton University Press, 1994.

Martinson, Ryan D., "The Arming of China's Maritime Frontier," *China Maritime Report*, No. 2, Newport, R.I.: U.S. Naval War College, June 2017. As of May 19, 2018: https://daisukybiendong.files.wordpress.com/2017/09/ryan-d-martinson-2017-the-arming-of-china_s-maritime-frontier-signed.pdf

Mazarr, Michael J., *Mastering the Gray Zone: Understanding a Changing Era of Conflict*, Carlisle Barracks, Pa.: U.S. Army War College Press, 2015.

Mazarr, Michael J., Jonathan Blake, Abigail Casey, Tim McDonald, Stephanie Pezard, and Michael Spirtas, *Understanding the Emerging Era of International Competition: Theoretical and Historical Perspectives*, Santa Monica, Calif.: RAND Corporation, RR-2726-AF, 2018. As of September 15, 2018: https://www.rand.org/pubs/research_reports/RR2726.html

Mazarr, Michael J., Gian Gentile, Dan Madden, Stacie L. Pettyjohn, and Yvonne K. Crane, *The Korean Peninsula: Three Dangerous Scenarios*, Santa Monica, Calif.: RAND Corporation, PE-262-A, 2018. As of July 30, 2018: https://www.rand.org/pubs/perspectives/PE262.html

McNair, Cory, *Worldwide Social Network Users: eMarketer's Estimates and Forecast for 2016–2021*, New York: eMarketer, July 12, 2017. As of May 3, 2018: https://www.emarketer.com/Report/Worldwide-Social-Network-Users-eMarketers-Estimates-Forecast-20162021/2002081

Mearsheimer, John J., *The Tragedy of Great Power Politics*, New York: W. W. Norton, 2001.

Messera, Heather, Ronald Keys, John Castellaw, Robert Parker, Ann C. Phillips, Jonathan White, and Gerald Galloway, *Military Expert Panel Report: Sea Level Rise and the U.S. Military's Mission*, 2nd ed., Washington, D.C.: Center for Climate and Security, 2018.

Messner, J. J., Nate Haken, Patricia Taft, Ignatius Onyekwere, Hannah Blyth, Charles Fiertz, Christina Murphy, Amanda Quinn, and McKenzie Horwitz, *2018 Fragile States Index*, Washington, D.C.: Fund for Peace, 2018. As of May 25, 2018: http://fundforpeace.org/fsi/2018/04/24/fragile-states-index-2018-annual-report/

Ministry of Defence UK, *Strategic Trends Programme: Future Operating Environment 2035*, London, November 30, 2014. As of October 2, 2017: https://www.gov.uk/government/uploads/system/uploads/attachment_data/file/646821/20151203-FOE_35_final_v29_web.pdf

Modernisation and Strategic Planning Division, Australian Army Headquarters, *Future Land Warfare Report*, Canberra, 2014. As of October 25, 2017: https://www.army.gov.au/sites/g/files/net1846/f/flwr_web_b5_final.pdf

Mollman, Steve, "It's Typhoon Season in the South China Sea—and China's Fake Islands Could Be Washed Away," *Quartz*, August 1, 2016. As of March 4, 2018: https://qz.com/745511/international-law-isnt-the-most-powerful-threat-to-chinas-artificial-islands-in-the-south-china-sea-nature-is/

Morgan, Forrest E., and Raphael S. Cohen, *Military Trends and the Future of Warfare: The Changing Global Environment and Its Implications for the U.S. Air Force*, Santa Monica, Calif.: RAND Corporation, RR-2849/3, 2020. As of May 2020:
https://www.rand.org/pubs/research_reports/RR2849z3.html

Morton, Louis, "War Plan Orange: Evolution of a Strategy," *World Politics*, Vol. 11, No. 221, 1959, pp. 221–250.

Murray, Williamson R., and Allan R. Millett, eds., *Military Innovation During the Interwar Period*, Cambridge, UK: Cambridge University Press, 1998.

———, *A War to Be Won: Fighting the Second World War*, Cambridge, Mass.: Belknap Press, 2000.

National Counterintelligence and Security Center, *Foreign Economic Espionage in Cyberspace*, Washington, D.C.: Office of the Director of National Intelligence, 2018. As of August 7, 2018:
https://www.dni.gov/files/NCSC/documents/news/20180724-economic-espionage-pub.pdf

National Intelligence Council, *Paradox of Progress*, Washington, D.C., Global Trends main report, January 2017. As of October 25, 2017:
https://www.dni.gov/index.php/global-trends-home

Ochmanek, David, *Restoring U.S. Power Projection Capabilities: Responding to the 2018 National Defense Strategy*, Santa Monica, Calif.: RAND Corporation, PE-260-AF, 2018. As of August 22, 2018:
https://www.rand.org/pubs/perspectives/PE260.html

Ochmanek, David, Peter A. Wilson, Brenna Allen, John Speed Meyers, and Carter C. Price, *U.S. Military Capabilities and Forces for a Dangerous World: Rethinking the U.S. Approach to Force Planning*, Santa Monica, Calif.: RAND Corporation, RR-1782-RC, 2017. As of July 26, 2018:
https://www.rand.org/pubs/research_reports/RR1782.html

Office of the Press Secretary, White House, "U.S.-India Joint Strategic Vision for the Asia-Pacific and Indian Ocean Region," Washington, D.C., January 25, 2015. As of November 28, 2017:
https://obamawhitehouse.archives.gov/the-press-office/2015/01/25/us-india-joint-strategic-vision-asia-pacific-and-indian-ocean-region

Office of the Under Secretary of Defense for Acquisition and Sustainment and Office of the Deputy Assistant Secretary of Defense for Manufacturing and Industrial Base Policy, *Fiscal Year 2017 Annual Industrial Capabilities Report to Congress*, Washington, D.C.: U.S. Department of Defense, March 2018. As of July 25, 2018:
http://www.businessdefense.gov/Portals/51/Documents/Resources/2017%20AIC%20RTC%2005-17-2018%20-%20Public%20Release.pdf?ver=2018-05-17-224631-340

Pachauri, R. K., and L. A. Meyer, eds., *Climate Change 2014: Synthesis Report. Contribution of Working Groups I, II and III to the Fifth Assessment Report of the Intergovernmental Panel on Climate Change*, Geneva, Switzerland: Intergovernmental Panel on Climate Change, 2014.

Pew Research Center, "Political Polarization in the American Public," June 12, 2014. As of November 8, 2017:
http://www.people-press.org/2014/06/12/political-polarization-in-the-american-public/

Pezard, Stephanie, Abbie Tingstad, Kristin Van Abel, and Scott Stephenson, *Maintaining Arctic Cooperation with Russia: Planning for Regional Change in the Far North*, Santa Monica, Calif.: RAND Corporation, RR-1731-RC, 2017. As of January 9, 2018:
https://www.rand.org/pubs/research_reports/RR1731.html

Posen, Barry R., *The Sources of Military Doctrine: France, Britain, and Germany Between the World Wars*, Ithaca, N.Y.: Cornell University Press, 1984.

President of the Russian Federation, *Strategiya natsional'noi bezopasnosti Rossiiskoi Federatsii [National Security Strategy of the Russian Federation]*, Moscow, Decree No. 683, December 31, 2015. As of January 3, 2018:
http://kremlin.ru/acts/bank/40391

———, "Presentation of Era Innovation Technopolis," Moscow: Kremlin, February 23, 2018. As of May 30, 2018:
http://en.kremlin.ru/events/president/news/56923

President's National Infrastructure Advisory Council, *Securing Cyber Assets: Addressing Urgent Cyber Threats to Critical Infrastructure*, Washington, D.C.: Department of Homeland Security, August 2017. As of August 7, 2018:
https://www.dhs.gov/sites/default/files/publications/niac-cyber-study-draft-report-08-15-17-508.pdf

Press, Daryl G., "The Myth of Air Power in the Persian Gulf War and the Future of Warfare," *International Security*, Vol. 26, No. 2, Fall 2001, pp. 5–44.

Radin, Andrew, and Clint Reach, *Russian Views of the International Order*, Santa Monica, Calif.: RAND Corporation, RR-1826-OSD, 2017. As of December 12, 2018:
https://www.rand.org/pubs/research_reports/RR1826.html

Rauhala, Emily, "Duterte Renounces U.S., Declares Philippines Will Embrace China," *Washington Post*, October 20, 2016. As of November 20, 2017:
https://www.washingtonpost.com/world/philippines-duterte-saysgoodbye-washington-and-helloto-beijing/2016/10/20/865f3cd0-9571-11e6-9cae-2a3574e296a6_story.html?utm_term=.6d91b967c317

Reisinger, Heidi, and Aleksandr Golts, *Russia's Hybrid Warfare: Waging War Below the Radar of Traditional Collective Defence*, Rome: NATO Defense College, Research Paper No. 105, November 2014. As of February 17, 2017:
https://www.files.ethz.ch/isn/185744/rp_105.pdf

Robinson, Linda, Todd C. Helmus, Raphael S. Cohen, Alireza Nader, Andrew Radin, Madeline Magnuson, and Katya Migacheva, *Modern Political Warfare: Current Practices and Possible Responses*, Santa Monica, Calif.: RAND Corporation, RR-1772-A, 2017. As of February 14, 2019:
https://www.rand.org/pubs/research_reports/RR1772.html

Roosevelt, Franklin D., "Fireside Chat," radio address, December 29, 1940, transcript via American Presidency Project. As of July 30, 2018:
http://www.presidency.ucsb.edu/ws/index.php?pid=15917

Rosen, Stephen Peter, "New Ways of War: Understanding Military Innovation," *International Security*, Vol. 13, No. 1, Summer 1988, pp. 134–168.

Rumsfeld, Donald H., "Transforming the Military," *Foreign Affairs*, Vol. 81, No. 3, May–June 2002, pp. 20–32.

Sarkees, Meredith Reid, "The COW Typology of War: Defining and Categorizing Wars (Version 4 of the Data)," paper, undated, as hosted on Meredith Reid Sarkees and Frank Wayman, *Resort to War: 1816–2007*, Washington, D.C.: CQ Press, 2010. As of October 24, 2017:
http://www.correlatesofwar.org/data-sets/COW-war

Shatz, Howard J., and Nathan Chandler, *Global Economic Trends and the Future of Warfare: The Changing Global Environment and Its Implications for the U.S. Air Force*, Santa Monica, Calif.: RAND Corporation, RR-2849/4, 2020. As of May 2020:
https://www.rand.org/pubs/research_reports/RR2849z4.html

Shpigel, Noa, "Friction in North Not Over After Extensive Syria Strikes, Israeli Defense Minister Warns," *Haaretz*, May 11, 2018. As of February 14, 2019:
https://www.haaretz.com/israel-news/
friction-not-over-after-extensive-syria-strikes-defense-chief-warns-1.6076202

Specia, Megan, "How Syria's Death Toll Is Lost in the Fog of War," *New York Times*, April 13, 2018. As of June 29, 2018:
https://www.nytimes.com/2018/04/13/world/middleeast/syria-death-toll.html

Statista, "Number of Social Network Users Worldwide from 2010 to 2021 (in billions)," webpage, undated. As of May 3, 2018:
https://www.statista.com/statistics/278414/
number-of-worldwide-social-network-users/

Stolfi, R. H. S., "Equipment for Victory in France in 1940," *History*, Vol. 55, No. 183, February 1970, pp. 1–20.

Tian, Nan, Aude Fleurant, Alexandra Kuimova, Pieter D. Wezeman, and Siemon T. Wezeman, "Trends in World Military Expenditure, 2017," SIPRI Fact Sheet, May 2018. As of June 29, 2018:
https://reliefweb.int/sites/reliefweb.int/files/resources/sipri_fs_1805_milex_2017.pdf

United Nations, Department of Economic and Social Affairs, Population Division, *The World's Cities in 2016—Data Booklet*, New York, 2016. As of February 18, 2019:
http://www.un.org/en/development/desa/population/publications/pdf/
urbanization/the_worlds_cities_in_2016_data_booklet.pdf

U.S. Air Force, *USAF Strategic Master Plan*, Washington, D.C., May 2015. As of July 26, 2018:
http://www.af.mil/Portals/1/documents/Force%20Management/Strategic_Master_Plan.pdf

U.S. Congressional Budget Office, *The Budget and Economic Outlook: 2018 to 2028*, Washington, D.C., April 2018.

U.S. Department of Defense, *Quadrennial Defense Review Report*, Washington, D.C., February 6, 2006. As of November 9, 2018:
http://archive.defense.gov/pubs/pdfs/qdr20060203.pdf

———, *Summary of the 2018 National Defense Strategy of the United States of America: Sharpening the American Military's Competitive Edge*, Washington, D.C., 2018. As of June 29, 2018:
https://www.defense.gov/Portals/1/Documents/pubs/2018-National-Defense-Strategy-Summary.pdf

U.S. Department of the Interior, Office of the Secretary, "Draft List of Critical Minerals," *Federal Register*, Vol. 83, No. 33, February 16, 2018, pp. 7065–7068. As of July 1, 2018:
https://www.federalregister.gov/documents/2018/02/16/2018-03219/draft-list-of-critical-minerals

U.S. Energy Information Administration, *Annual Energy Outlook 2017*, Washington, D.C.: U.S. Department of Energy, January 5, 2017. As of July 1, 2018:
https://www.eia.gov/outlooks/archive/aeo17/pdf/0383(2017).pdf

U.S. Marine Corps, Futures Directorate, *2015 Marine Corps Future Security Environment Forecast: Futures 2030–2045*, Quantico, Va., 2015. As of October 2, 2017:
http://www.mcwl.marines.mil/Portals/34/Documents/2015%20MCSEF%20-%20Futures%202030-2045.pdf

Van Evera, Stephen, "The Cult of the Offensive and the Origins of the First World War," *International Security*, Vol. 9, No. 1, Summer 1984, pp. 58–107.

Victor, David G., "What Resource Wars?" *The National Interest*, November 12, 2007. As of July 24, 2018:
https://nationalinterest.org/article/what-resource-wars-1851

von Clausewitz, Carl, *On War*, ed. and trans., Michael Howard and Peter Paret, Princeton, N.J.: Princeton University Press, 1984.

VornDick, Wilson T., "Thanks Climate Change: Sea-Level Rise Could End South China Sea Spat," *The Diplomat*, November 8, 2012.

———, "China's Island Building + Climate Change: Bad News," *Real Clear Defense*, March 9, 2015. As of March 4, 2018:
https://www.realcleardefense.com/articles/2015/03/10/chinese_island_reclamation_the_climate_change_challenge_107722-2.html

Watts, Stephen, "Air War and Restraint: The Role of Public Opinion and Democracy," in Matthew Evangelista, Harald Muller, and Niklas Schorning, eds., *Democracy and Security: Preferences, Norms, and Policy-Making*, New York: Routledge, Taylor & Francis Group, 2008, pp. 56–60.

Weisgerber, Marcus, "The US Is Raiding Its Global Bomb Stockpiles to Fight ISIS," *Defense One*, May 26, 2016. As of August 22, 2018:
https://www.defenseone.com/threats/2016/05/us-raiding-its-global-bomb-stockpiles-fight-isis/128646/

White House, *National Security Strategy of the United States of America*, Washington, D.C., December 2017. As of July 2, 2018:
https://www.whitehouse.gov/wp-content/uploads/2017/12/NSS-Final-12-18-2017-0905.pdf

World Bank, *High and Dry: Climate Change, Water, and the Economy*, Washington, D.C., 2016. As of July 27, 2018:
http://www.worldbank.org/en/topic/water/publication/high-and-dry-climate-change-water-and-the-economy

———, "World Development Indicators," database, May 3, 2018. As of May 21, 2018:
https://datacatalog.worldbank.org/dataset/world-development-indicators

Work, Robert, Deputy Secretary of Defense, "The Third U.S. Offset Strategy and Its Implications for Partners and Allies," speech before the Center for a New American Security and the NATO Allied Command Transformation, Washington, D.C., January 28, 2015.